T0181146

The Information Retrieval Series   Volume 35

More information about this series at http://www.springer.com/series/6128

Massimo Melucci

Introduction to Information Retrieval and Quantum Mechanics

 Springer

Massimo Melucci
University of Padua
Padova, Italy

ISSN 1387-5264
The Information Retrieval Series
ISBN 978-3-662-56916-0   ISBN 978-3-662-48313-8   (eBook)
DOI 10.1007/978-3-662-48313-8

Springer Heidelberg New York Dordrecht London

Printed on acid-free paper

Springer-Verlag GmbH Berlin Heidelberg is part of Springer Science+Business Media
(www.springer.com)

To my family.

Preface

Information Retrieval (IR) is the complex of activities performed by a computer system so as to retrieve from a collection of documents all and only the documents which contain information relevant to the user's information need. The peculiar difficulty of IR is the fact that relevance cannot be precisely and exhaustively described using data; for example, a relevant document cannot be precisely and exhaustively described using text even if this text includes or considers all elements or aspects of a topic; a user's information need cannot be precisely and exhaustively described using a query even if this query is a fully comprehensive description of the need. An unbridgeable gap between relevance and document content and between information need and user's request exists, such that what can be said about the document's relevance or the user's information need can only be inferred by the document content and the user's request (or all the other sources of evidence available to an IR system).

Information retrieval researchers are constantly searching for theoretical frameworks which help them make a major breakthrough in the overall effectiveness of retrieval systems. Since its advent, information retrieval has been marked by a series of significant theoretical advances. It is customary to correspond these advances with the logical models, the vector space models, and the probabilistic models. More recently, the machine learning models were added to this series of advances. All these models are now not only the main pillars of modern systems and applications, they are also the building blocks of the formulation of new models and techniques, thanks to the mathematical frameworks of logic, geometry, and probability on which they rely. Recently, a new perspective from theoretical physics was added to this series of advances.

Newton's laws provide a correct description of physical systems and in particular macroscopic systems (e.g., sand grains, balls, and planets) in terms of, for example, position, size, or distance. However, these laws must be revised when position, size, or distance is observed at the microscopic level, thus making the prediction provided by these laws uncertain. In the nineteenth century, the gap between observation and prediction gave rise to Quantum Mechanics (QM), which deals with the mathematical description of the motion and interaction of subatomic particles.

QM established the impossibility of measuring physical systems at the microscopic level with arbitrary precision, thus legitimizing the axiom that physical systems at the microscopic level cannot be precisely and exhaustively observed using any device and that any observation can always be subjected to a probability measure of the degree to which the observed value is real. Therefore, there is an unbridgeable gap other than the gap mentioned as regards IR; it lies between the unknowable world of subatomic particles and the outcomes produced by the devices used for describing this world.

The similarity between the barrier across the space separating relevance from content, or information need from request, and the barrier across the space separating the values observed through a device from the reality of subatomic particle was the reason why some researchers investigated the utilization in IR of the quantum mechanical framework. The fundamental idea underlying this utilization was the potential offered by the quantum mechanical framework to predict the values which can only be observed in conditions of uncertainty. However, this uncertainty is not only caused by the existence of different observable values (e.g., the number of spots on a dice), but it is also caused by the ignorance of the internal structure of the system under observation. The ignorance of the internal structure of the user's background, context, and plans of future actions causes an uncertainty different from the uncertainty caused by the variety of possible values that can be observed from the user. The first uncertainty is caused by the interaction between the measurement of the observable variables that describe the user's background, context, and plans and the user's background, context, and plans themselves. When these aspects are measured, not only some variable values are observed, but these values come to existence and evolve because of the measurement devices. The second uncertainty is "only" caused by the fact that there is more than one possible value that can be observed, and the actual outcome of a measurement depends on the randomness intrinsic in the measurement device; for example, the outcome of the draw of a dice depends on the randomness of the draw, and the uncertainty is caused by the existence of six distinct values.

In the first decade of this century, some studies investigated whether the quantum mechanical framework could be applied to research areas such as human cognition, natural language processing, and information retrieval other than theoretical physics. These studies and the results thereof can be regarded as an approach to investigating these research areas that might be named as "quantum inspired" or "quantum like." The idea behind the quantum-like approach to disciplines other than physics is that, although the quantum properties exhibited by particles such as photons cannot be exhibited by macroscopic objects, some phenomena can be described by the language or have some characteristics of the phenomena (e.g., superposition or entanglement) described by the quantum mechanical framework in physics. The other idea is that because modeling information processing may be a very daunting task, the designer of an information system such as an information retrieval system needs to be provided with theoretical frameworks that go beyond the traditional logic, geometric, or probabilistic frameworks.

The main thrust of this book is to illustrate how the quantum mechanical framework has been and may be applied to IR. The book is placed at the intersection between IR and QM which leverages the similarity between the barrier across the space separating relevance from content, or information need from request, and the barrier across the space separating the values observed through a device from the reality of subatomic particles. The book is not another proposal of retrieval model; in contrast, it aims to highlight the correspondences between the physical phenomena observed through the quantum mechanical framework (e.g., entanglement, superposition, and interference) and the phenomena which might be encountered during the retrieval process. The book is not about quantum phenomena in IR; in contrast, it aims to propose the use of the mathematical language of the quantum mechanical framework for describing the mode of operation of a retrieval system. The book cannot be an exhaustive description of the potential offered by the quantum mechanical framework; however, it aims to motivate researchers to delve into this framework and find further correspondences between IR and QM and design effective retrieval and indexing procedures which leverage the potential of the framework. To this end, the book is organized in four chapters.

Chapter 1 illustrates the main modeling approaches to information retrieval. This chapter is not an exhaustive illustration of the retrieval models proposed in decades of research; in contrast, it aims to highlight the main concepts which will be further addressed in the other two chapters of the book. In particular, the chapter illustrates Boolean logic, vector spaces, and two probabilistic models because these three concepts occur and are integrated in the quantum mechanical framework; it also briefly describes the machine learning-based approach to information retrieval since it has been a significant approach and may open further research directions within the intersection with the quantum mechanical framework. Although the first chapter is about information retrieval, it also introduces some notions of the quantum mechanical framework to build the bridge between the two fields.

Chapter 2 briefly explains the main concepts of the quantum mechanical framework; the literature is immense both because of the intrinsic complexity and the long time that has passed since the advent of the subject; therefore, we selected the concepts that may be linked to information retrieval; for example, superposition is explained in this chapter because it was investigated in some research reported in the third chapter. Some topics—entanglement is a glaring example—may appear rather difficult and sometimes really arduous. The book introduces them as gently as possible and suggests some further and more technical readings.

Chapter 3 surveys the research conducted in the intersection between information retrieval and the quantum mechanical framework. The chapter is organized in topics, and each topic is often ascribable to one research group or a few research groups. The literature on the intersection between information retrieval and the quantum mechanical framework has often been reported in journal or conference papers; only a small part has been reported as a book. Each chapter ends with a section suggesting the most relevant and interesting readings and listing the books and papers by which this book has been inspired and prepared.

Chapter 4 concludes with some suggestions for future research. Some research directions that are considered essential topics are briefly outlined in the hope that the quantum mechanical framework will be fully leveraged to achieve effective and efficient information retrieval systems.

The approach taken by this book is to organize the presentation of the main notions of the quantum mechanical framework (e.g., incompatibility, interference, superposition, and entanglement) in terms of observables and probability as it is customarily done in information retrieval in order to explain how they can inspire a new view and give rise to a new potential of information retrieval. Indeed, information retrieval systems are based on probability and observables in a similar way as incompatibility, interference, superposition, and entanglement are. The illustration based on probability and observables is not constrictive, since a large part of the literature of the quantum mechanical framework has this emphasis. On the contrary, this illustration leverages the long tradition in probability and statistics and naturally fits with many foundational problems of information retrieval.

This book is intended to be accessible to computer scientists in general and in particular to researchers working in information retrieval, database systems, and machine learning. The reader is expected to be a postgraduate, a PhD student, or a researcher who wants to have a clear picture in a short time of the potential of the quantum mechanical framework to pursue his own research interests. The mention of some crucial topics of information retrieval, the introduction to the main notions of the quantum mechanical framework, and the illustration of the way the quantum mechanical framework has inspired information retrieval are indeed intended to link the reader's background in information retrieval with the new notions he is acquiring while reading this book. The introduction of some crucial topics of information retrieval also aims to make this book accessible to computer scientists without a strong background in information retrieval and willing to address the use of the quantum mechanical framework in fields other than physics from the standpoint of an information scientist. Finally, the book may be of interest to noncomputer scientists who have utilized the quantum mechanical framework as an inspiration for their research carried out in domains other than physics.

The introduction of the quantum mechanical framework inevitably utilizes mathematical instruments; these instruments are based on the complex vector spaces which are the theoretical infrastructure that support the probability measures of the uncertainty occurring in QM and the logic underlying the observables applied to the physical world. The use of mathematical instruments may be problematic since they make the comprehension of the quantum mechanical framework harder, and, importantly, they might be used to build theoretical descriptions with no correspondence to the physical world; indeed, Polkinghorne (2002) and Zeilinger (2010) highlighted that some mathematical dissertations can lead to misleading or erroneous descriptions of the physical world. Such a mistake can be avoided only if the real world of IR is carefully examined.

As the topic described in this book has a relatively short history, the notation and the definitions used in the publications surveyed in this book may be inconsistent. Therefore, a common notation and glossary for notations and definitions are used in this book to make the understanding of the notions of the quantum mechanical framework less demanding than would otherwise be the case if these notions were studied using different publications.

Although almost every researcher acknowledges the importance of theory in information retrieval, they are reluctant to adopt further mathematical formalisms without clear evidence that this adoption brings about some actual experimental improvements. Because the use of the quantum mechanical framework in information retrieval is based on the mathematical formalism of the Hilbert spaces and has been debated for some years without reaching a consensus, the main expected benefit gained from this book is the clarification of whether, why, and when the use of the mathematical formalism and of the notions of the quantum mechanical framework can effectively be adopted for addressing some foundational issues of information retrieval.

Padova, Italy   Massimo Melucci
June 2015

Acknowledgments

I am indebted to Keith van Rijsbergen for his seminal scientific contribution to the topic of this book. I would like to also thank Diederik Aerts, Maristella Agosti, Gianfranco Cariolaro, Andrei Khrennikov, and Giuseppe "Pino" Vallone and the referees for their help.

Thank you, Alessandra and Oleg, for your support.

Contents

Chapter 1
Elements of Information Retrieval

As the main aim of the book is to illustrate the intersection between information retrieval and the quantum mechanical framework, this chapter illustrates those concepts of information retrieval which can be intersected with the framework. In particular, the main notions of the most important modeling approaches to designing and implementing information retrieval systems are explained in this chapter before they are revisited, generalized, and extended within the quantum mechanical framework in the following chapters. After introducing the core concepts of information retrieval, we introduce the Boolean model and logic, the vector space model, the main probabilistic models, and briefly the machine learning approach to ranking documents. The chapter ends with some suggestions for further reading.

1.1 Introduction

IR is nowadays a key technology for collecting, organizing, and providing access to information. It is therefore not surprising that the basic concepts of this technology involve the notion of information. However, as this is a technology at the user's service, the notions of information need and relevance also play a crucial role. Given the abstract nature of information, need and relevance, an IR system has to utilize data to implement and make these abstract notions usable and, from a computer science point of view, both effective and efficient. In this section, we briefly illustrate these notions.

In the most essential respects, IR is about information. Although its origin may be simply explained by the Latin words *in* (into) and *forma* (shape), thus suggesting that information shapes the internals of an organization, information is a very complex notion of which many definitions have been provided in different research and disciplinary fields. Despite this complexity, the notion of information may be simplified when it is viewed from the point of view of a user of a computer system.

© Springer-Verlag Berlin Heidelberg 2015
M. Melucci, *Introduction to Information Retrieval and Quantum Mechanics*,
The Information Retrieval Series 35, DOI 10.1007/978-3-662-48313-8_1

For the aims of an IR system, information is intended as whatever changes the user's knowledge.

The role played by information in forming the user's knowledge will be insufficient unless it is provided with an effective means to make information processing possible. To be effective, information needs to be represented by symbols or signals to be conveyed to the users; these symbols and signals are called "data." The most important data managed by an IR system are documents and queries.

Documents are the basic artifacts managed by an IR system. A document is a data container provided with an identifier that is independent of the content. Although a document can be rendered in a way appropriate to make information usable or more simply agreeable, the data contained in a document are supposed to be unstructured or "flat." Therefore, no structured model can be applied to the data of a document in the same way as the relational model is used to describe a database. The absence of any data structure makes the management of these data difficult and a source of problems for any system. The other artifact managed by a system is the collection, that is, a container of documents gathered through an algorithm which can be manually, semiautomatically, or fully automatically performed, the latter being the most common mechanism used by the search engines of the World Wide Web (WWW).

In most cases, documents contain only written text. Although this type of data is less problematic than other media such as image, video, and sound, text is still affected by the ambiguity of natural language, thus making textual IR difficult. The most common causes of ambiguity are polysemy and synonymy: the former is the property of words that have more than one meaning; the latter is the property of pairs of words that have the same meaning. These two main causes of ambiguity can be exacerbated by multilingualism when collections, documents, or queries contain words from different natural languages.

Queries are the most utilized data that represent the user's information needs. A query is essentially a sentence expressed in a natural language; it may be very short (e.g., one word) or much longer (e.g., a text paragraph). Despite being a widespread means to represent information needs, it is not the only means; other means can be, for example, the data observed during the interaction between the user and the system or the data about preferences collected through a social network.

Suppose a user has an information need originated from a problem and represented by queries or other data—a user has a problem to solve and needs information to find the solution. Not all the information available in the data is useful or necessary; on the contrary, in a small amount, data represent useful or necessary information to solve the user's problem. The property of information that meets a user's information need is called "relevance." As both users and problems vary according to context, relevance is context dependent, and therefore, what is deemed relevant to one user and his problem may no longer be relevant to another user or for solving another problem.

To be effective and make prediction feasible, current IR technology requires that a retrieval model be defined and tested before any scientific evaluation. An IR model is a set of algebraic structures that describes documents and queries. A model defines an operation called retrieval function that maps these information structures to the numeric real field. The most effective models are based on Boolean logic (Sect. 1.2), vector spaces (Sect. 1.3), and probability theory from which the relevance model and the language models have been derived (Sect. 1.4). The machine learning approach to IR (Sect. 1.5) aims to find the retrieval function looking at the data collected during the interaction between the users and the system.

Therefore, IR is the complex of activities that represents information as data and retrieves data that represents information relevant to the user's information needs.

From an architectural point of view, an IR system is a computer system performing IR activities. The main activities are indexing and retrieval. Indexing is concerned with extracting and organizing document content descriptors, that is, data about data and sometimes called metadata; in the event of text, a content descriptor is a term (e.g., a keyword). Content descriptors are organized in one or more indexes, that is, data structures consisting of lists of postings where a posting is a data structure that relates a term with a document, and therefore, a posting list is a data structure that relates a term with documents. Retrieval consists of processing the user's query (or other data that represent an information need) and of matching the user's query with the documents stored in the indexes. The result of matching query and documents can be ranked according to some retrieval function implementing a retrieval model based on a modeling approach.

By its nature, IR is inherently an interactive activity performed by a user accessing the collections managed by a system. Through such an interaction, the user aims to refine his query, to provide additional evidence describing his information need, or to indirectly tell his needs to the system. Although the literature offers a myriad of options for implementing user interaction, the best known methodology that supports this interaction is Relevance Feedback (RF).

RF is the process that updates information need descriptions using additional data observed during the user interaction with the system. RF may be explicit when the additional data are explicitly provided by the user as depicted in Fig. 1.1a; the user's data are usually scores, comments, or marks given to the retrieved documents and stored in the system for further processing also of queries other than the query with respect to which the assessments are provided. RF may also be pseudo when the additional data are collected from the documents deemed relevant by the system as depicted in Fig. 1.1b. "Pseudo" originates from Greek and means "falsehood"; when applied to feedback, "pseudo" means that relevance is not the true, real relevance provided by a user, it is on the contrary a relevance provided by a surrogate for the user, i.e., the system. Finally, RF may be implicit when the additional data are

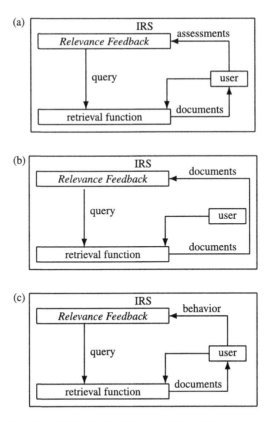

Fig. 1.1 Relevance Feedback. (**a**) Explicit. (**b**) Pseudo. (**c**) Implicit

collected from the user's behavior as depicted in Fig. 1.1c. "Implicit" means that relevance is implied by data other than the true, real relevance provided by a user who explicitly tells the system what are the relevant documents.

RF is illustrated in Fig. 1.2 which depicts a system consisting of a user, a RF module, and a retrieval function. Initially, there is a system in which the user participates, a retrieval function and a feedback function (Fig. 1.2a). Then, the user sends a query to the retrieval function which returns a list of retrieved documents (Fig. 1.2b). Afterward, the user or the system assesses the relevance of the retrieved documents and sends the assessments to the feedback function (Fig. 1.2c). Finally, the feedback function sends the feedback data (e.g., a new query) to the retrieval function which returns a new list of documents to the user (Fig. 1.2d).

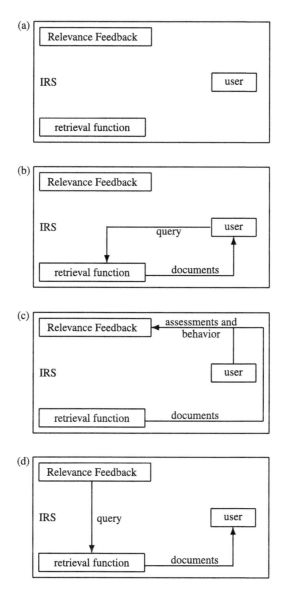

Fig. 1.2 Steps of Relevance Feedback

1.2 Boolean Logic

In this section, we briefly introduce Boolean logic in IR, because it is the main bridge between the complex of notions of IR and the quantum mechanical framework. Indeed, these two fields share the use of some basic logical notions such as event, proposition, and operator for expressing more complicated concepts. What distinguishes the quantum mechanical framework from the framework in which IR is defined is given by some generalizations of the properties of the mathematical objects used to represent those basic logical notions.

Although its development dates back to the early days of IR systems and may appear to be an old-fashioned technology, nowadays Boolean logic is also frequently used by those modern systems that rank the documents retrieved to answer the user's queries using complex scoring functions; for example, the search engines of the WWW utilize this logic to filter the pages before they are ranked using many different statistical sources of evidence as described in Sect. 1.5.

1.2.1 Sets and Operators

The success of the application of Boolean logic in IR derives from the fact that it relies on the naïve set theory. This theory views a content descriptor as a set of documents and then a document as an element of a set. The effectiveness of Boolean logic is due to the very natural view of a document collection as a set of documents held by the end users. Thanks to this view, a user expects to receive a set of documents as the answer to his query.

Although the user's query may be very complex, the power of Boolean logic in modeling the user's queries stems from the very important equivalence between the Boolean operators and the operators of set theory. The operation of union of two sets corresponds to the disjunction operator, the operator of intersection of two sets corresponds to the conjunction operator, and the operation of difference between two sets corresponds to the negation operator. According to this correspondence, the user can formulate queries with varying complexity knowing that the use of an operator induces a specific set operation; for example, he knows that the set of documents returned by a query including a conjunction operator is not greater than the set returned by a query including a disjunction operator.

Not only the user's queries are Boolean expressions where the operands are content descriptors and the query operators are the logical operators. Complex content descriptors can be defined as Boolean expressions too, thus making indexing algorithms able to build "intelligent" descriptions of the informative content of the documents; for example, an expression like "apple AND banana AND NOT cherry" can describe the informative content of a document that is about both apples and bananas but not about cherries in a more informative way than three single and independent content descriptors like "apple," "banana," and "cherry." An example is depicted in Fig. 1.3.

Boolean logic has been the most used model in early IR systems, thanks to its effectiveness in retrieving large proportions of documents relevant to many information needs and at the same time only small proportions of nonrelevant documents. Nevertheless, it has been adopted by a few search engines of the WWW since it is little loved by most end users who are asked to devote much interaction and to have expertise both in the topic of their information need and in the use

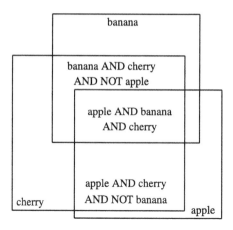

Fig. 1.3 Documents, queries, and Boolean logic

of the Boolean logic thereof. This logic may cause confusion since the meaning of the logic operators might be obscure, and there may be more than one Boolean expression for expressing a query which would more effectively be expressed using natural language.

Other drawbacks may be caused by the retrieved document set which may be very large or very small up to two extreme cases: null output when the Boolean expression of the query is too specific or output overload when the Boolean expression of the query is too generic. An approach to overcoming the drawbacks of Boolean modeling has been the coordination level, which lets propositions (such as queries) made of atomic propositions have nonbinary values, thus ranking documents over the real line, whereas the binary values (i.e., 0 and 1) of Boolean modeling without coordination level cannot permit the ranking of documents.

1.2.2 Coordination Level

To overcome the problems of null output and output overload, the notion of coordination level was introduced. The coordination level is a measure of the degree to which each returned document matches the query. In this way, the coordination level provides a score for ranking the documents. This ranking allows the user to decide how many documents to inspect and allows the system to cut the bottom-ranked documents off the list.

The coordination level is calculated as follows. A Boolean query is transcribed in Conjunctive Normal Form (CNF), namely, as a list of propositions related by the conjunction operator; each of these propositions are disjunctions of atomic propositions. The coordination level of a document is the number of propositions composing a CNF satisfied by the document. Suppose, for example, A, B, C are the document subsets associated to three index terms such as "apple," "banana," and "cherry," respectively; let $A = \{d_1, d_2\}$, $B = \{d_2, d_3\}$, $C = \{d_3\}$, and $A \wedge (C \vee B)$ be the query. It follows that the coordination level of d_1 is 1 because only the first proposition is satisfied by d_1, that is, d_1 only belongs to A; the coordination level of d_2 is 2 because d_2 belongs to both A and $C \vee B$; the coordination level of d_3 is 1 because it only belongs to $C \vee B$. Note that d_1, d_3 would not have been retrieved if the coordination level had not been calculated because they do not satisfy the query; d_1 does not belong to $C \vee B$ nor d_3 belongs to A.

A variation of the coordination level was introduced to take the variable size of document subsets into account; indeed, the document subsets are of arbitrary size, and therefore, a small subset may be treated as a large subset. This variation has been called weighted coordination level: instead of assigning a constant weight to each proposition made true by a document, a different weight is assigned depending on the proposition; for example, an Inverse Document Frequency (IDF) may be used. The weighted coordination level is then the sum of the weights assigned to each proposition.

At the query processing, an IR system leverages the CNF of a query. The disjunctions are processed to build k posting lists. For each disjunction $T_{j,1} \vee \cdots \vee T_{j,n_j}$, the n_j posting lists that correspond to $T_{j,1}, \ldots, T_{j,n_j}$ are retrieved and merged. The k posting lists that result from these merging operations are then processed; each of these posting lists corresponds to a disjunction and is conjoined by a series of k conjunction operators (i.e., $\wedge$ or AND). CNF query processing is efficient because (i) a document must occur in every posting list built after merging the posting lists of the disjunctions and (ii) the IR system processes the k posting lists in ascending order of size, that is, from the least frequent term to the most frequent term. In this way, the system can skip over the posting list of frequent terms to find the documents that also contain the infrequent terms.

1.2.3   Weight Functions

The weight functions of the Boolean logic are introduced in this section since they play an important role in the link between this logic and the projectors of the vector spaces introduced in the following section. Algebraically, the Boolean logic consists of weight functions mapping an element (e.g., a document) to a numeric value which is usually chosen from the set $\{0, 1\}$. A weight function can be defined as

$$w_A : S \to \{0, 1\}$$

where S is the set of elements (e.g., a collection of documents) and $A \subseteq S$ is a subset of elements corresponding to, for example, a term; for example, A may be the subset of documents indexed by a term A. Let x be a document and A be a term. The proposition that x is indexed by the term, that is, $x \in A$, is either true or false. This formulation is equivalent to $w_A(x) = 1$ if and only if $x \in A$. We have a binary

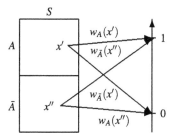

Fig. 1.4 Function mapping set elements to a real number

function when $w_A(x) \in \{0, 1\}$, yet it may be a nonbinary function when $w_A(x) \in \mathbb{R}$. The function w can also be called a weight function for terms because it provides the weight of A in x; an example is depicted in Fig. 1.4. For each Boolean operator (i.e., negation, conjunction, and disjunction), a weight function can be defined as follows.

1.2.3.1   Weight Function of the Negation Operator

The weight function for the negation operator can be defined as follows. Let x be a document and $\bar{A}$ be the subset of documents that make a negation of A true. A binary weight function for the negation operator is

$$w_{\bar{A}}(x) = 1 - w_A(x)$$

as depicted in Fig. 1.4.

1.2.3.2   Weight Function of the Conjunction Operator

Consider the weight function for the conjunction operator (i.e., AND). To this end, let x be a document and $C = A \wedge B$ be a conjunction of two subsets. It follows that $w_C(x)$ is either 1 or 0, i.e., $x \in C$ is either true or false. In particular,

$$w_C(x) = 1 \text{ if and only if } w_A(x) = 1 \text{ and } w_B(x) = 1$$

as depicted in Fig. 1.5. A weight function that follows the definition of the conjunction above is

$$w_C(x) = w_A(x)w_B(x)$$

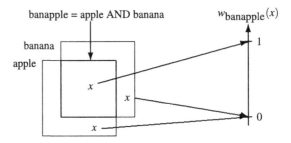

Fig. 1.5 Weight function of a conjunction

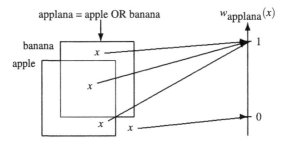

Fig. 1.6 Weight function of a disjunction

1.2.3.3   Weight Function of the Disjunction Operator

Consider the weight function for the disjunction operator (OR). Let x be a document and $D = A \vee B$ be a disjunction. The proposition $x \in D$ is either true or false in correspondence with the fact that $w_D(x)$ is either 1 or 0. In particular,

$$w_D(x) = 1 \text{ if and only if } w_A(x) = 1 \text{ or } w_B(x) = 1$$

A weight function that follows the definition of the disjunction above is

$$w_D(x) = w_A(x) + w_B(x) - w_A(x)w_B(x)$$

as depicted in Fig. 1.6.

1.2.4   *Dirac Notation and Projectors*

At this point of the explanation of the use of Boolean logic in IR, it is necessary to introduce the formulation of the weight function using the mathematical concepts of the quantum mechanical framework, i.e., vectors, spaces, and operators thereof. The notation used to represent the quantum mechanical framework is different from that usually adopted in the IR literature and was introduced by Dirac (1935); before proceeding with the explanation, a digression on this notation is therefore necessary.

1.2.4.1 Dirac Notation

According to the Dirac notation, a vector x of the d-dimensional complex vector space H is written as

$$|x\rangle = \begin{pmatrix} x_1 \\ x_2 \\ \vdots \\ x_d \end{pmatrix}$$

and is called "ket." The conjugate transpose of $|x\rangle$ is represented as

$$\langle x| = \left(x_1^*, x_2^*, \ldots, x_d^*\right)$$

and is called "bra"[1] where x_i^* is the conjugate of the complex number x_i. The inner product between x and y is represented as $\langle x|y\rangle$, which is a complex number since the vectors are defined over the complex vector space H. In particular, suppose

$$|x\rangle = \begin{pmatrix} x_1 \\ x_2 \\ \vdots \\ x_d \end{pmatrix} \qquad |y\rangle = \begin{pmatrix} y_1 \\ y_2 \\ \vdots \\ y_d \end{pmatrix}$$

The inner product is defined as

$$\langle x|y\rangle = \sum_{i=1}^{d} x_i^* y_i$$

The outer product (or dyad) is written as

$$|x\rangle\langle y| = \begin{pmatrix} x_1 y_1^* & \cdots & x_1 y_d^* \\ & \ddots & \\ x_d y_1^* & \cdots & x_d y_d^* \end{pmatrix}$$

---

[1] This is the reason why the Dirac notation is also called the bra(c)ket notation.

If $\mathbf{A}$ is a matrix (or an operator), then $\mathbf{A}|x\rangle$ is the vector resulting from the linear transformation represented by $\mathbf{A}$. In particular, suppose

$$|x\rangle = \begin{pmatrix} x_1 \\ x_2 \\ \vdots \\ x_d \end{pmatrix} \qquad \mathbf{A} = \begin{pmatrix} a_{1,1} & \cdots & a_{1,d} \\ & \ddots & \\ a_{d,1} & \cdots & a_{d,d} \end{pmatrix}$$

The result of the application of the operator on the vector can be written as

$$\mathbf{A}|x\rangle = \begin{pmatrix} \sum_{i=1}^{d} a_{1,i} x_i \\ \sum_{i=1}^{d} a_{2,i} x_i \\ \vdots \\ \sum_{i=1}^{d} a_{d,i} x_i \end{pmatrix}$$

1.2.4.2 Projectors

A projector $\mathbf{A}$ is an operator (i.e., a function) that maps a vector to itself; this is the reason why one speaks of projection. Suppose A is a subset of documents which corresponds to a term. In mathematical terms, a projection can be written as follows:

$$\mathbf{A}|x\rangle = |x\rangle \text{ if and only if } \langle x|\mathbf{A}|x\rangle = 1 \text{ if and only if } |x\rangle \in A$$

Therefore, the projector is a function that tests whether a document x belongs to A, that is, the document is indexed by the term. It can be shown that a projector has the following property:

$$\mathbf{A}\mathbf{A} = \mathbf{A}$$

Note that a projector can be represented as a matrix. A pictorial description of projection is provided in Fig. 1.7.

A special case of a projector is provided by a vector of unitary length. Each vector corresponds to one projector defined as the outer product between the (column) vector and its conjugate transpose (i.e., the row vector). To explain this, suppose $|x\rangle$ is a vector; the projector corresponding to this vector is a special dyad or outer product written as

$$|x\rangle\langle x|$$

To check that it is indeed a projector, it is necessary and sufficient to compute the square of the projector and find that it results in the projector itself, as follows:

$$(|x\rangle\langle x|)(|x\rangle\langle x|) = |x\rangle(\langle x|x\rangle)\langle x| = (\langle x|x\rangle)|x\rangle\langle x| = |x\rangle\langle x|$$

since $\langle x|x\rangle = 1$.

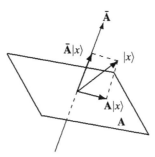

Fig. 1.7 Projector of a subspace

A projector determines a vector subspace including all the vectors that are mapped to itself when the projector is applied to them. Thanks to this correspondence between projector and subspace, a subspace of H can be viewed as a subset of vectors where the projector plays the role of the mechanism that checks whether a vector belongs to the subspace.

The other correspondence is between subspace and subset; from an IR point of view, a mathematical correspondence is not actually of interest, although a conceptual one is. The conceptual correspondence is given by the possibility of expressing a proposition like "this element belongs to this subset" in terms of projection, that is, it is possible to say that "this element belongs to this subset" when the vector corresponding to the element is mapped on itself.

The crucial fact is that there exists one projector for each subspace. Given the correspondences between subspace, projector, and proposition, we can say that the mathematical object used to represent propositions is the projector.

The correspondence between the quantum mechanical framework and the weight functions can be summarized by the following table:

Set	S	Complex vector space	H
Subset	A	Subspace	A
Set element	x	Vector	$\lvert x \rangle$
Weight function	w_A	Projector	$\mathbf{A}$
Membership	$w_A(x) = 1$	Projection	$\mathbf{A}\lvert x \rangle = \lvert x \rangle$
Proposition	$x \in A$	Projection	$\mathbf{A}\lvert x \rangle = \lvert x \rangle$

1.2.4.3 Negation, Mutual Exclusiveness, and Resolution to Unity

After having explained how projectors can represent propositions, this section explains how the Boolean operators (i.e., conjunction, disjunction, and negation) can be expressed using functions of projectors, these functions essentially being matrix products and sums. Using projectors, negation can be represented by

$$\bar{\mathbf{A}} = \mathbf{1} - \mathbf{A}$$

where $\mathbf{1}$ is the identity projector such that

$$\mathbf{1}|x\rangle = |x\rangle \text{ for all } |x\rangle \in H$$

that is, the $d \times d$ identity matrix

$$\mathbf{1} = \begin{pmatrix} 1 & \cdots & 0 \\ & \ddots & \\ 0 & \cdots & 1 \end{pmatrix}$$

The identity projector corresponds to the proposition $|x\rangle \in S$, which is always true for all elements x or equivalently for all vectors $|x\rangle$ of H. The operator $\bar{\mathbf{A}}$ is still a projector since

$$\bar{\mathbf{A}}\bar{\mathbf{A}} = \bar{\mathbf{A}}(\mathbf{1} - \mathbf{A}) = \bar{\mathbf{A}} - (\mathbf{1} - \mathbf{A})\mathbf{A} = \bar{\mathbf{A}} - (\mathbf{A} - \mathbf{A}) = \bar{\mathbf{A}}$$

An example of projector that corresponds to a negation (i.e., complement of a subset) is depicted in Fig. 1.7.

Consider, for example, the projectors of two mutually exclusive events $A, \bar{A}$, say "relevant" and "not relevant." As this event space is binary, the projectors corresponding to the propositions are defined in the bidimensional space and can be written as follows:

$$\mathbf{A} = \frac{1}{2}\begin{pmatrix} 1 & 1 \\ 1 & 1 \end{pmatrix} \qquad \bar{\mathbf{A}} = \frac{1}{2}\begin{pmatrix} 1 & -1 \\ -1 & 1 \end{pmatrix}$$

The quantum mechanical formulation of the mutual exclusiveness of the events $A, \bar{A}$ can be represented by the orthogonality between the projectors, that is,

$$\mathbf{A}\bar{\mathbf{A}} = \mathbf{0}$$

where $\mathbf{0}$ represents the impossible event.

In a similar way, the property that these events (or propositions) are all the possible propositions about an event is represented by the resolution to unity of the projectors, that is,

$$\mathbf{A} + \bar{\mathbf{A}} = \mathbf{1}$$

where $\mathbf{1}$ is the identity projector.

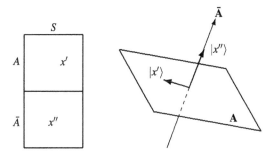

Fig. 1.8 Elements and sets in a vector space

When the quantum mechanical framework is applied to IR, a document x is viewed as a unit vector, that is, $\langle x|x \rangle = 1$; the set S of elements is viewed as a complex vector space, and the subset of documents A is viewed as a subspace. Figure 1.8 depicts an element which corresponds to a vector in a subspace and a subset which corresponds to a subspace; the subspaces are named after their projectors as the subspaces are in one-to-one correspondence with the projectors, that is, only one projector $\mathbf{A}$ can be defined for this subspace.

1.2.4.4 Conjunction and Commutativity

To represent conjunction using vectors and projectors, we have to use three projectors $\mathbf{A}$, $\mathbf{B}$, and $\mathbf{C}$ corresponding to three events A, B, and C, respectively, where

$$C = A \wedge B$$

Suppose $\mathbf{A}$ and $\mathbf{B}$ commute with respect to the matrix product, that is, the following equivalence holds:

$$\mathbf{C} = \mathbf{AB} = \mathbf{BA}$$

If $|x\rangle \in C$, then

$$\langle x|\mathbf{C}|x \rangle = \langle x|\mathbf{AB}|x \rangle = 1$$

only if $\mathbf{A}|x\rangle$ and $\mathbf{B}|x\rangle$ are the same vector $|x\rangle$, that is, if and only if

$$\mathbf{A}|x\rangle = |x\rangle \qquad \mathbf{B}|x\rangle = |x\rangle$$

On the other hand, when these two conditions hold, we have that

$$\mathbf{A}|x\rangle = \mathbf{AB}|x\rangle = |x\rangle \qquad \mathbf{B}|x\rangle = \mathbf{BA}|x\rangle = |x\rangle$$

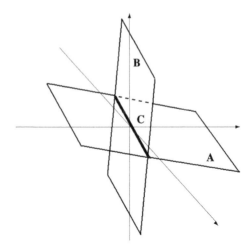

Fig. 1.9 Projectors and conjunction

and

$$\mathbf{AB} = \mathbf{BA} = \mathbf{C}$$

The operator $\mathbf{C}$ is still a projector since

$$\mathbf{CC} = (\mathbf{AB})(\mathbf{AB}) = \mathbf{A}(\mathbf{BA})\mathbf{B} = \mathbf{A}(\mathbf{AB})\mathbf{B} = (\mathbf{AA})(\mathbf{BB}) = \mathbf{AB} = \mathbf{C}$$

An example is depicted in Fig. 1.9; the subspace in common to the subspaces corresponding to the projectors $\mathbf{A}$ and $\mathbf{B}$ corresponds to a projector $\mathbf{C}$ and is a line. In the event that $\mathbf{A}$ does not commute with $\mathbf{B}$, one speaks about incompatibility between observables or propositions. Incompatibility is a point of radical departure of the quantum mechanical framework from the classical framework upon which IR relies. Incompatibility is introduced in Sect. 2.2.2.

1.2.4.5 Disjunction and Span

To represent disjunction using vectors and projectors, we have two projectors $\mathbf{A}$ and $\mathbf{B}$ and a third projector $\mathbf{D}$ corresponding to A, B, and D, respectively, where

$$D = (A \vee B) \setminus C$$

Suppose

$$\mathbf{D} = \mathbf{A} + \mathbf{B} - \mathbf{C}$$

If $|x\rangle \in A$ or $|x\rangle \in B$, then

$$\langle x|\mathbf{D}|x\rangle = \langle x|\mathbf{A} + \mathbf{B} - \mathbf{C}|x\rangle = 1$$

The operator $\mathbf{D}$ is still a projector since

$$
\begin{aligned}
\mathbf{DD} &= \mathbf{AD} + \mathbf{BD} - \mathbf{CD} \\
&= \mathbf{A}(\mathbf{A} + \mathbf{B} - \mathbf{C}) + \mathbf{B}(\mathbf{A} + \mathbf{B} - \mathbf{C}) - \mathbf{C}(\mathbf{A} + \mathbf{B} - \mathbf{C}) \\
&= \mathbf{AA} + \mathbf{AB} - \mathbf{AC} + \mathbf{BA} + \mathbf{BB} - \mathbf{BC} - \mathbf{CA} - \mathbf{CB} + \mathbf{CC} \\
&= \mathbf{A} + \mathbf{AB} - \mathbf{AAB} + \mathbf{AB} + \mathbf{B} - \mathbf{BBA} - \mathbf{BAA} - \mathbf{ABB} + \mathbf{AB} \\
&= \mathbf{A} + \mathbf{AB} - \mathbf{AB} + \mathbf{AB} + \mathbf{B} - \mathbf{AB} - \mathbf{AB} - \mathbf{AB} + \mathbf{AB} \\
&= \mathbf{A} + \mathbf{B} - \mathbf{AB} \\
&= \mathbf{A} + \mathbf{B} - \mathbf{C} \\
&= \mathbf{D}
\end{aligned}
$$

Figure 1.10 depicts an example of how disjunction is implemented by two projectors corresponding to two orthogonal rays. The resulting subspace is a plane. It can be noted that $\mathbf{D}$ mirrors a vector $|x'\rangle$ on itself if this vector mirrors on $\mathbf{B}$ and mirrors a vector $|x''\rangle$ on itself if this vector mirrors on $\mathbf{A}$ as expected. Indeed, x' belongs to B and x'' belongs to A. It follows that both x' and x'' belong to $A \cup B$. Consider a document x''' other than x' and x'' which does not belong to A nor to B. Following the guidelines, x''' corresponds to a vector $|x'''\rangle$ in the vector space. This vector can arbitrarily be chosen; we place it in the subspace of $\mathbf{D}$.

What is not expected is that $\mathbf{D}$ mirrors a vector $|x'''\rangle$ on itself although this vector is not mirrored on $\mathbf{A}$ or $\mathbf{B}$. Suppose a vector $|x'''\rangle$ belongs to the subspace spanned by $|x'\rangle$ and $|x''\rangle$, that is,

$$|x'''\rangle = a'|x'\rangle + a''|x''\rangle \qquad |a'|^2 + |a''|^2 = 1$$

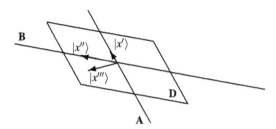

Fig. 1.10 Projectors and disjunction

This vector does not belong to the subspace determined by $\mathbf{A}$ nor does it belong to the subspace determined by $\mathbf{B}$ since

$$\mathbf{A}|x'''\rangle = a'\mathbf{A}|x'\rangle + a''\mathbf{A}|x''\rangle = a'\mathbf{A}|x'\rangle = a'|x'\rangle \neq |x'''\rangle$$

and

$$\mathbf{B}|x'''\rangle = a'\mathbf{B}|x'\rangle + a''\mathbf{B}|x''\rangle = a''\mathbf{B}|x''\rangle = a''|x''\rangle \neq |x'''\rangle$$

However, $|x'''\rangle$ belongs to the subspace determined by $\mathbf{D}$ since

$$
\begin{aligned}
\mathbf{D}|x'''\rangle &= (\mathbf{A} + \mathbf{B} - \mathbf{AB})(a'\mathbf{A}|x'\rangle + a''\mathbf{A}|x''\rangle) \\
&= a'\mathbf{A}|x'\rangle + a'\mathbf{B}|x'\rangle - a'\mathbf{AB}|x'\rangle + a''\mathbf{A}|x''\rangle + a''\mathbf{B}|x''\rangle - a''\mathbf{AB}|x''\rangle \\
&= a'|x'\rangle + a''|x''\rangle \\
&= |x'''\rangle
\end{aligned}
$$

Therefore, $|x'''\rangle$ belongs to the disjunction of two subspaces although it does not belong to either subspace. This is a signal that the logic induced by the vector spaces is more general than the logic induced by sets. This generalization happens because $|x'\rangle$ and $|x''\rangle$ span a subspace larger than the simple union of the subspaces spanned by $|x'\rangle$ and $|x''\rangle$.

1.3 Vector Space Model

The early formulation of the Vector Space Model (VSM) was by Salton (1963) who further developed the model in the 1970s. The VSM was then experimented and applied to several tasks in the 1980s until its deployment in actual systems (e.g., the search engines of the WWW) in the 1990s.

In this section, the VSM is introduced because it is hinged on some concepts of the theory of the vector spaces in common with the quantum mechanical framework and the Boolean logic introduced in Sect. 1.2. It is therefore worth highlighting that the VSM can be integrated with Boolean logic by leveraging the common concepts of the quantum mechanical framework. Moreover, in this way, the quantum mechanical framework can be applied to IR and there generalized since these concepts are already utilized by the VSM and Boolean logic.

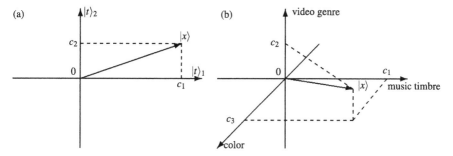

Fig. 1.11 Intuition of the Vector Space Model

1.3.1   Basis and Indexing

The conceptual idea underlying the VSM is that a content descriptor is a basis vector. It follows that an index managed by a system corresponds to the basis of a vector space over the real field. The number of distinct descriptors and then the size of the index are equal to the dimension of the vector space. On the other hand, documents are vectors spanned by the basis vectors, that is, each document vector is a weighted linear combination of the vector of the basis corresponding to an index. Similarly, queries are vectors spanned by the same basis. In general, any data contained in information objects managed by a system can be represented as a vector.

The intuition behind this model is illustrated in Fig. 1.11a. A user starts interacting with the system without any terms in mind; this point corresponds to the origin $(0, 0)$ of a coordinate system. Once the user selects a term t_1, the point moves to $|t_1\rangle$ with weight or coordinate c_1, and the information object can be represented by $c_1|t_1\rangle$. When the user chooses $|t_2\rangle$ with weight or coordinate c_2, the vector of the information object is $c_1|t_1\rangle + c_2|t_2\rangle$. When three terms are selected, the final result is given by

$$|x\rangle = c_1|t_1\rangle + c_2|t_2\rangle + c_3|t_3\rangle$$

The weights or coordinates measure the importance of a term in the information object or in the collection of objects. The same applies to multimedia objects where the content descriptors can be video genre, music timbre, or color as depicted in Fig. 1.11b.

The VSM is based on a few notions of the theory of vector spaces. The first notion is that of linear independence. Let

$$\{|t_1\rangle, \ldots, |t_k\rangle\}$$

be a set of vectors of the real d-dimensional vector space $\mathbb{R}^d$ where $k \leq d$. The set $\{|t_1\rangle, \ldots, |t_k\rangle\}$ is linearly independent when

$$|x\rangle = \sum_{i=1}^{k} c_i |t_i\rangle = \sum_{i=1}^{k} b_i |t_i\rangle$$

implies

$$c_i = b_i \qquad i = 1, \ldots, k$$

This means that no vector in $\{|t_1\rangle, \ldots, |t_k\rangle\}$ can be a linear combination of the remaining vectors. When it is linearly independent, the set $\{|t_1\rangle, \ldots, |t_k\rangle\}$ is a basis of dimension k of the vector space defined over $\mathbb{R}^d$.

From an IR point of view, independence provides an important feature of the VSM. Independence means that the set of terms included in an index managed by a system are all useful for defining the vectors representing the information objects in a collection. Term dependence will be exploited by Latent Semantic Analysis (LSA) to compute the hidden factors that determine how terms are used in a document collection as explained in Sect. 1.3.5.

For any vector $|x\rangle$ of the k-dimensional space spanned by $\{|t_1\rangle, \ldots, |t_k\rangle\}$,

$$|x\rangle = \sum_{i=1}^{k} c_i |t_i\rangle \qquad x_j = \sum_{i=1}^{k} c_i t_{ij} \qquad j = 1, \ldots, d$$

where t_{ij} is the j-th component of $|t_i\rangle$ and $k \leq d$.

In general, a basis is not necessarily canonical nor orthogonal; it is only an independent set of vectors. A basis is orthogonal when $\langle t_i | t_j \rangle = 0$ for each $i \neq j$. It follows that when

$$|x\rangle = c_1 |t_1\rangle + \cdots + c_k |t_k\rangle$$

then

$$c_i = \langle t_i | x \rangle$$

An instance of an orthogonal basis is the canonical basis, that is,

$$|t_1\rangle = \begin{pmatrix} 1 \\ 0 \\ \vdots \\ 0 \end{pmatrix} \qquad |t_2\rangle = \begin{pmatrix} 0 \\ 1 \\ \vdots \\ 0 \end{pmatrix} \qquad \cdots \qquad |t_k\rangle = \begin{pmatrix} 0 \\ 0 \\ \vdots \\ 1 \end{pmatrix}$$

where $|t_i\rangle$ has null elements except for the i-th element which is 1.

1.3.2   Inner Product and Retrieval

The retrieval function of the VSM maps a pair document-query to a real number. As documents and queries are vectors, the most used retrieval function is based on the inner product between vectors (Fig. 1.12).

The retrieval function of the VSM is based on the inner product between two vectors. In the VSM, the vector space is defined over the real field; therefore, the conjugate of a scalar is the scalar itself, and the conjugate transpose of a vector is the transpose vector. It follows that

$$\langle x|y \rangle = \sum_{j=1}^{d} x_j y_j$$

where $\langle x|$ is a row vector; an example of inner product is depicted in Fig. 1.11 which explains that $\langle x|y \rangle = 2 \times (-2) + 1 \times (-2) = -6$ is the size of the projection of $|y\rangle$ on the ray spanned by $|x\rangle$ multiplied by the cosine of the angle between the vectors and by the size of $|x\rangle$, that is,

$$\langle x|y \rangle = \|x\| \|y\| \cos \theta$$

When the vectors are unit vectors, the inner product is just the cosine of the angle; otherwise, the vector sizes measure some properties of the information objects such as document length. Note that the cosine may be negative, while most applications of the VSM do not consider this possibility. Moreover, it is assumed that the model is defined over the real field—in contrast, the quantum mechanical framework is

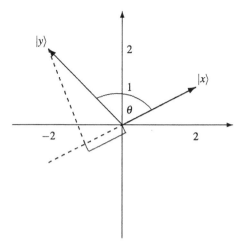

Fig. 1.12 Inner product in the real bidimensional space

defined over the complex field, and therefore, notions like "cosine of the angle" do not make much sense.

1.3.3 Connections with Boolean Logic

It is possible now to make a connection with the description of Boolean logic given in terms of projectors in Sect. 1.2 and therefore with the quantum mechanical framework. This connection consists of corresponding an orthogonal basis to a collection of projectors such that

$$\mathbf{A}_i = |t_i\rangle\langle t_i| \qquad i = 1, \ldots, k$$

where $\mathbf{A}_i$ is a projector. It can be easily checked that the projectors are mutually orthogonal. Using the resolution to unity, it can be seen that

$$\mathbf{A}_1 + \cdots + \mathbf{A}_k = \mathbf{1}$$

where $\mathbf{1}$ is the projector of the k-dimensional vector space spanned by the basis. It follows that a vector defined according to the VSM such as

$$|x\rangle = c_1|t_1\rangle + \cdots + c_k|t_k\rangle \tag{1.1}$$

can be viewed as the result of the linear combination of k propositions since a projector corresponds to a proposition and a basis vector corresponds to a projector. In particular, the projector $\mathbf{A}_i$ can multiply both sides of (1.1) to obtain

$$\mathbf{A}_i|x\rangle = c_1\mathbf{A}_i|t_1\rangle + \cdots + c_k\mathbf{A}_i|t_k\rangle$$

As the basis vectors are mutually orthogonal,

$$\mathbf{A}_i|t_j\rangle = \begin{cases} \mathbf{0} & i \neq j \\ |t_j\rangle & i = j \end{cases}$$

It follows that

$$\mathbf{A}_i|x\rangle = c_i\mathbf{A}_i|t_i\rangle = c_i|t_i\rangle$$

and that

$$\langle t_i|\mathbf{A}_i|x\rangle = \langle t_i|x\rangle = c_i$$

Therefore, a coefficient c_i can be viewed as the "answer" to the question, for example, whether a document represented by $|x\rangle$ is about the term represented by $|t_i\rangle$. This coefficient calculated as

$$c_i = \langle x|t_i\rangle$$

may in principle be any real number. If c_i were binary, the answer to the question on the aboutness of the document would be either "yes" or "no," that is, $c_i = 1$ if and only if the document is about the term; otherwise, $c_i = 0$.

It may be that more than one coefficient can be different from 0, thus obtaining a document vector $|x\rangle$ which represents a document about more than one term at the same time; for example,

$$|x\rangle = c_1|t_1\rangle + c_2|t_2\rangle$$

represents a document about t_1 and t_2. Orthogonality plays a significant role since, were the basis vectors not orthogonal, the question whether the document is about, say, t_1 would produce a number different from c_1, in particular

$$\langle t_1|x\rangle = c_1 + c_2\langle t_1|t_2\rangle$$

thus representing a situation where the aboutness of the document to t_1 depends on the aboutness to t_2 and to a measure of the relationship between t_1 and t_2, this measure being given by $\langle t_1|t_2\rangle$.

We have highlighted above some connections between the sets of Boolean logic and the vector spaces of the VSM, in particular, how a weight function w_A can be expressed as a projector $\mathbf{A}$ to test the membership of an element to set A. As a projector is an operator acting on vector spaces, it is natural to see connections with the VSM. There are some differences, however.

Using Boolean logic, a query A, $A \wedge B$, or $A \vee B$ can be expressed as a projector $\mathbf{A}$, $\mathbf{AB}$, or $\mathbf{A} + \mathbf{B} - \mathbf{AB}$, respectively. A document vector $|x\rangle$ either belongs to the subspace of the query projector or does not.

Using the VSM, although a query y is represented as a vector $|y\rangle$, it can correspond to a projector $|y\rangle\langle y|$, but it cannot be easily expressed as a Boolean expression such as $\mathbf{A}$, $\mathbf{AB}$, or $\mathbf{A} + \mathbf{B} - \mathbf{AB}$.

In general, the projector $|y\rangle\langle y|$ describes a subspace placed obliquely to the subspaces described by the projectors of the single terms. In particular, when the basis is orthogonal, a query vector can be written as

$$|y\rangle = \langle t_1|y\rangle|t_1\rangle + \cdots + \langle t_k|y\rangle|t_k\rangle$$

where $\langle t_i|y\rangle$ is the size of the projection of $|y\rangle$ on $|t_i\rangle$. When a projector $|t_i\rangle\langle t_i|$ is assigned to the basis vector $|t_i\rangle$, it is obtained that

$$\langle y|t_i\rangle\langle t_i|y\rangle = |\langle t_i|y\rangle|^2$$

As $|t_i\rangle\langle t_i|$ corresponds to a term according to the use of the Boolean logic, we have that $|y\rangle$ is a weighted linear combination of term vectors where the weights of this combination measure the degree to which the i-th term describes the query. Moreover, the inner product used in the VSM becomes

$$\langle x|y\rangle = \langle t_1|y\rangle\langle x|t_1\rangle + \cdots + \langle t_k|y\rangle\langle x|t_k\rangle$$

thus highlighting that it is a weighted sum of measures of the degree to which the document is described by a term.

One issue is how to implement the coefficients $\langle t_i|y\rangle$ and $\langle x|t_i\rangle$ since the meaning of any real number is less clear than the meaning of a binary value, which denotes either "yes" or "no." Some direction has been provided by the coordination level which measures the membership of an element to a subset, yet it is rather heuristic, and a more principled mechanism would allow the assignment of these coefficients to be governed. Although a mechanism that provides a value to these coefficients is described in Sect. 1.3.4, a meaning can only be provided when probability comes into play as illustrated in Sect. 1.4.

1.3.4 Weighting Schemes

The coefficients of the linear combinations of the document vectors and query are defined by means of formulas called "weighting schemes." The main schemes are known as binary, Term Frequency (TF), IDF, and TF × IDF (TFIDF).

The binary scheme of the weight of descriptor j in object (e.g., document) i is very simple:

$$c_{ij} = \begin{cases} 1 \text{ if } t_j \text{ occurs in } i \\ 0 \text{ otherwise} \end{cases}$$

As the binary scheme equally treats the objects independently of descriptor, the TF scheme can be defined as follows. Let f_{ij} be the frequency of term j in document i:

$$c_{ij} = f_{ij}$$

However, the TF scheme equally treats the objects independently of whether the descriptors are common or rare, the latter being a feature helping to identify the relevant documents.

To this end, the IDF scheme is defined as follows. Let $n_j > 0$ be the number of documents indexed by term j and let N be the total number of documents:

$$c_{ij} = \log N/n_j$$

The TFIDF combines the previous two schemes as follows. Let f_{ij} be the frequency of term j in document i, let n_j be the number of documents indexed by term j, and let N be the total number of documents:

$$c_{ij} = f_{ij} \log \frac{N}{n_j}$$

1.3.5   Latent Semantic Analysis

LSA has been one of the most significant contributions in understanding the potential of the vector spaces in IR. Initially proposed to cope with the problem of synonymy and polysemy, it has been further studied and applied in other tasks and problems as an alternative approach to the use of thesauri and dictionaries.

Based on the well-known Singular Value Decomposition (SVD) theorem, LSA allows for simultaneously modeling term and document relationships. It does not need very special input; it needs only a term-document occurrence matrix. To start, consider, for example, two terms and one document represented in the bidimensional space as follows:

$$|x_1\rangle = \begin{pmatrix} \frac{1}{\sqrt{2}} \\ \frac{1}{\sqrt{2}} \end{pmatrix}$$

using the canonical basis

$$|t_1\rangle = \begin{pmatrix} 1 \\ 0 \end{pmatrix} \qquad |t_2\rangle = \begin{pmatrix} 0 \\ 1 \end{pmatrix}$$

This example means that whenever $|t_1\rangle$ is used with weight c_1, $|t_2\rangle$ is used with the same weight; this situation is depicted in Fig. 1.13. Recall that

$$|x_1\rangle = \frac{1}{\sqrt{2}} \begin{pmatrix} 1 \\ 0 \end{pmatrix} + \frac{1}{\sqrt{2}} \begin{pmatrix} 0 \\ 1 \end{pmatrix} = c_{11} \begin{pmatrix} 1 \\ 0 \end{pmatrix} + c_{12} \begin{pmatrix} 0 \\ 1 \end{pmatrix}$$

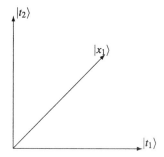

Fig. 1.13 Document vector spanned by the canonical basis

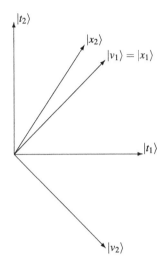

Fig. 1.14 Document vector spanned by an alternative basis

and that the canonical basis vectors are orthonormal, that is, the terms are uncorrelated since vector orthogonality corresponds to incorrelation, and the occurrence of a term does not influence the occurrence of the other term. As $|t_1\rangle$ are $|t_2\rangle$ are used with the same constant weight, their use is perfectly correlated. Since there is perfect correlation, it would be possible to use one term only and predict the other term with certainty; these equal weights indeed mean that it is certain that the use of $|t_1\rangle$ within $|x_1\rangle$ with weight c_1 implies that $|t_2\rangle$ is used within $|x_1\rangle$ with the same weight; note that this correlation is between the terms within a given document. If we replace $|t_1\rangle$, $|t_2\rangle$ with $|v_1\rangle$, we obtain Fig. 1.14. Now let us define the "negation" of $|v_1\rangle$ and name it $|v_2\rangle$, thus obtaining the following vector basis:

$$|v_1\rangle = \begin{pmatrix} +\frac{1}{\sqrt{2}} \\ +\frac{1}{\sqrt{2}} \end{pmatrix} \qquad |v_2\rangle = \begin{pmatrix} +\frac{1}{\sqrt{2}} \\ -\frac{1}{\sqrt{2}} \end{pmatrix}$$

The basis vectors $|v_1\rangle$, $|v_2\rangle$ represent terms more complex than the "canonical" terms $|t_1\rangle$, $|t_2\rangle$ and are given by the co-occurrence in the same document $|x_1\rangle$. The orthogonality of $|v_1\rangle$, $|v_2\rangle$ may represent some term relationships; $|v_2\rangle$ is a term opposite to $|v_1\rangle$ since $\langle v_1|v_2\rangle = 0$. Moreover, note that $|v_1\rangle$, $|v_2\rangle$ is an orthogonal basis alternative to $|t_1\rangle$, $|t_2\rangle$. Therefore, the document vector can be expressed using this alternative basis as follows:

$$|x_1\rangle = 1|v_1\rangle + 0|v_2\rangle = |v_1\rangle$$

A similar result can be obtained when there are two documents. Suppose

$$|x_2\rangle = \begin{pmatrix} \frac{2}{\sqrt{13}} \\ \frac{3}{\sqrt{13}} \end{pmatrix} = \frac{2}{\sqrt{13}}|t_1\rangle + \frac{3}{\sqrt{13}}|t_2\rangle$$

The alternative basis can still be used to express this document as follows:

$$|x_2\rangle = \begin{pmatrix} +\frac{5}{\sqrt{26}} \\ -\frac{1}{\sqrt{26}} \end{pmatrix} = \frac{5}{\sqrt{26}}|v_1\rangle - \frac{1}{\sqrt{26}}|v_2\rangle \qquad (1.2)$$

The coefficients of the latter expression are the sizes of the projection of $|x_2\rangle$ on the lines spanned by $|v_1\rangle$ and $|v_2\rangle$ and the projection is

$$|z_2\rangle = \langle v_1|x_2\rangle|v_1\rangle = \frac{5}{\sqrt{26}}|v_1\rangle$$

where $\langle v_1|x_2\rangle$ is the size of the projection of $|x_2\rangle$ on $|v_1\rangle$ and is obtained by multiplying (1.2) by $\langle v_1|$.

When the number of documents increases, the sum of the sizes of the projections of the document vectors on the basis vectors should be kept as large as possible for a given basis. Given n documents (or objects) represented by $|x_1\rangle, \ldots, |x_n\rangle$ in $\mathbb{R}^d$, the $d \times n$ occurrence matrix $\mathbf{X}$ can be prepared. Suppose $|z_i\rangle$ is the projection of $|x_i\rangle$ on the q-dimensional subspace spanned by the basis $|v_1\rangle, \ldots, |v_q\rangle$ where $(q < d)$. As the projection is a vector expressed in terms of $|v_i\rangle$,

$$|z_i\rangle = \langle v_1|x_i\rangle|v_1\rangle + \cdots + \langle v_q|x_i\rangle|v_q\rangle$$

where $\langle v_j|x_i\rangle$ is the size of the projection of $|x_i\rangle$ on $|v_j\rangle$. When $q = 1$, one basis vector is sought to obtain

$$\langle v_1|\mathbf{X} = \big(\langle v_1|x_1\rangle \cdots \langle v_1|x_n\rangle\big)$$

This alternative vector $|v_1\rangle$ is chosen to maximize the overall projection of the $|x_i\rangle$ on $|v_1\rangle$, that is,

$$|v_1\rangle = \arg\max_{|v\rangle} \|\langle v|\mathbf{X}\|^2 - \sigma(\langle v|v\rangle - 1)$$

It can be shown that $|v_1\rangle$ is the main eigenvector of the term correlation matrix $\mathbf{XX}^*$; for example, consider the following matrix:

$$\mathbf{X} = \begin{pmatrix} \frac{1}{\sqrt{2}} & \frac{2}{\sqrt{13}} \\ \frac{1}{\sqrt{2}} & \frac{3}{\sqrt{13}} \end{pmatrix}$$

The correlation matrix is as follows:

$$\mathbf{XX}^* = \begin{pmatrix} \frac{21}{26} & \frac{25}{26} \\ \frac{25}{26} & \frac{31}{26} \end{pmatrix} \approx \begin{pmatrix} 0.808 & 0.962 \\ 0.962 & 1.192 \end{pmatrix}$$

The following eigenvectors can be computed:

$$\mathbf{V} = \begin{pmatrix} 0.634 & 0.773 \\ 0.773 & -0.634 \end{pmatrix} = (|v_1\rangle \; |v_2\rangle)$$

to obtain the projections:

$$\langle v_1|x_1\rangle = 0.99 \quad \langle v_1|x_2\rangle = 0.99 \quad \langle v_2|x_1\rangle = 0.01 \quad \langle v_2|x_2\rangle = 0.01$$

Thus, we have seen that

$$\mathbf{XX}^*|v_i\rangle = \sigma_i|v_i\rangle \qquad i = 1, \ldots, r$$

where r is the rank. Similarly, the eigenvectors of $\mathbf{X}^*\mathbf{X}$ are

$$\mathbf{X}^*\mathbf{X}|u_i\rangle = \eta_i|u_i\rangle \qquad i = 1, \ldots, r$$

In general, the SVD can be computed:

$$\underset{d \times n}{\mathbf{X}} = \underset{d \times d}{\mathbf{V}} \; \underset{d \times n}{\Sigma} \; \underset{n \times n}{\mathbf{U}^*}$$

that can be written as follows:

$$\mathbf{X} = \sigma_1|u_1\rangle\langle v_1| + \cdots + \sigma_r|u_r\rangle\langle v_r|$$

When applied to the context of IR, the truncated SVD is often used since it allows us to obtain a reduced vector space preserving the correlations between terms:

$$\mathbf{X}_q = \sigma_1|u_1\rangle\langle v_1| + \cdots + \sigma_q|u_q\rangle\langle v_q|$$

where $q < r$ or even $q \ll r$ when r is very large as is customary in IR. SVD aims at minimizing the Frobenius norm defined as follows:

$$\|\mathbf{A}\|_F = \left(\sum_{i,j} |a_{i,j}|^2 \right)^{\frac{1}{2}}$$

Therefore, the optimal truncated SVD minimizes:

$$||\mathbf{X} - \mathbf{X}_q||_F = \sqrt{\sigma_{q+1} + \cdots \sigma_k}$$

As a consequence, occurrence, co-occurrence, and correlation matrices are approximated by $\mathbf{X}_q$.

An early application of LSA was for document retrieval. An example of the use of LSA in finding relevant documents begins with the following occurrence matrix:

$$\mathbf{X} = \begin{pmatrix} +1 & +2 & +1 & +2 \\ +1 & +3 & -2 & -1 \end{pmatrix}$$

Each column refers to a document and each row refers to a term; it follows that four document vectors are depicted in the bidimensional space spanned by the canonical basis (Fig. 1.15). The four documents (the column vectors of the occurrence matrix) are assigned the following relevance assessments:

$$\begin{pmatrix} 1 & 1 & 0 & 0 \end{pmatrix}$$

so that the document vectors in the first quadrant of Fig. 1.15 are relevant and the document vectors in the fourth quadrant are nonrelevant. Consider a query represented by the following vector:

$$|y\rangle = \begin{pmatrix} 1 \\ 0 \end{pmatrix}$$

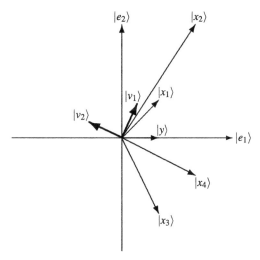

Fig. 1.15 Four document vectors in the bidimensional space

The vector $|y\rangle$ represents an ambiguous query because it is between the cluster of relevant documents in the first quadrant and the cluster of nonrelevant documents in the fourth quadrant. Indeed, it provides the following document ranking:

$$\langle y|X = (1\ 2\ 1\ 2)$$

A loss of precision and loss of recall can be observed because the first term is ambiguous. Suppose we compute the SVD of X^*:

$$X^* = V\Sigma U^*$$

thus obtaining the following decomposition:

$$V = \begin{pmatrix} 0.42 & 0.91 \\ 0.91 & -0.42 \end{pmatrix}$$

$$\Sigma = \begin{pmatrix} 4.1 & 0 & 0 & 0 \\ 0 & 2.9 & 0 & 0 \end{pmatrix}$$

$$U = \begin{pmatrix} 0.328 & 0.164 & -0.930 & 0.032 \\ 0.880 & 0.184 & 0.334 & -0.283 \\ -0.342 & 0.598 & -0.040 & -0.723 \\ -0.014 & 0.762 & 0.151 & 0.629 \end{pmatrix}$$

Note that $|v_1\rangle$ spans a ray passing close to the relevant documents, whereas $|v_2\rangle$ spans a ray passing close to the nonrelevant documents. One expects that the inner product between relevant document vectors (i.e., $|x_1\rangle$ and $|x_2\rangle$) and the first eigenvector $|v_1\rangle$ will be high and positive, whereas the inner product between nonrelevant document vectors (i.e., $|x_3\rangle$ and $|x_4\rangle$) and the second eigenvector $|v_2\rangle$ will be high and negative. The document vectors can be projected on the subspace spanned by $|v_1\rangle$ to obtain

$$\langle v_1|X = (1.330\ 3.565\ -1.387\ -0.057)$$

Since the relevant documents have the highest scores, both recall and precision can be maximized.

1.3.6   Relevance Feedback

According to the cluster hypothesis, the relevant documents tend to resemble relevant documents more than nonrelevant documents. If the documents were depicted in a bidimensional space, it may be possible to see when the cluster

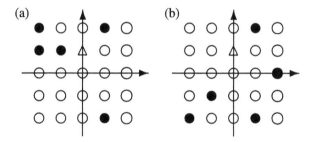

Fig. 1.16 When the cluster hypothesis holds and when it does not

hypothesis holds (see Fig. 1.16a where the relevant documents, which are depicted as black circles, are next to each other), when it does not (see Fig. 1.16b where the nonrelevant documents are dispersed among the relevant documents), and how the retrieval performed to answer a query (represented by a triangle) is affected.

The cluster hypothesis is important both because of efficiency reasons and effectiveness reasons. When the cluster hypothesis holds, the relevant documents can be stored together and retrieved with less access operations than those necessary when the relevant documents are not clustered; the number of access operations would be proportional to the number of clusters. Moreover, the retrieved clusters would be dense with relevant documents.

RF leverages the cluster hypothesis. Suppose a query is represented as follows:

$$|y\rangle = \sum_{i=1}^{k} b_i |t_i\rangle$$

Given r relevant documents, $d_1, \ldots, d_r$, and $n - r$ nonrelevant documents, $d_{r+1}, \ldots, d_n$, the modified query would be represented as follows:

$$|y\rangle' = |y\rangle + \sum_{j=1}^{r} \alpha_j |d_j\rangle + \sum_{h=r+1}^{n} \beta_h |d_h\rangle \qquad \alpha_j \geq 0 \qquad \beta_h \leq 0$$

The first summation of the right-hand side is called positive RF, whereas the second summation is called negative RF. How these two mechanisms operate is depicted in Fig. 1.17. Suppose there are five relevant documents in a collection of 25 documents (top-left of the figure). When a query is sent to the IR system, the VSM puts the query close to both nonrelevant and relevant documents (top-right). After positive RF, the query moves closer to the relevant documents (bottom-left) and further moves closer after negative RF (bottom-right).

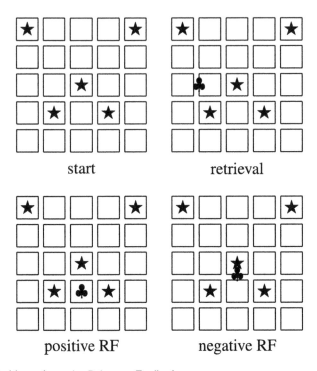

start   retrieval

positive RF   negative RF

Fig. 1.17 Positive and negative Relevance Feedback

1.4   Probabilistic Models

The probabilistic models of IR have in common the presence of a universe of elementary events where an event is a single occurrence of a process or phenomenon; it is called "elementary" since it cannot be decomposed into simpler occurrences. In the literature of probability theory, the notion of elementary event is distinguished from that of event; the latter is a subset of elementary events. The notion of event is crucial because a probability measure that maps an event to the unit range [0, 1] is applied to obtain the degree of belief that an elementary event belongs to the event.

The role played by the probabilistic models has become important since the Boolean model has been difficult to apply in IR tasks; the researcher has to cope with the lack of ranking, and the end user has to face null output and output overload. The VSM succeeded in improving the user's experience since it provides some ranking, but it leaves the problem of finding the coefficients of the linear combinations open.

The probabilistic models provide an answer to this problem yet at a principled level, that is, it explains how to provide the weights of coordination level used when Boolean logic is applied to IR and also provides an explanation of why the TFIDF of the VSM has been so effective. However, its impact is not only at a principled level—at present, the probabilistic models are also well accepted at the industrial level. In this section, we survey two important probabilistic models for IR: the relevance model and the language models.

1.4.1 Relevance Model

We call this model "relevance model" because it explicitly represents relevance. The most known implementation, which is illustrated in this section, is also known as "BM25" or "Okapi" or "Robertson-Sparck Jones" model. According to this model, the document collection is viewed as a universe of elementary events. The terms or in general the content descriptors resulting from indexing the collection determine document subsets and therefore implement the events of the probability space. The key notion of this model is that relevance is a document set A, and therefore, it is modeled as an event.

As these events are subsets of a probability space, a probability measure can be applied, and each event (i.e., terms, relevance and their subsets, and logical combinations) is assigned a probability; see also Fig. 1.18 which depicts the event "relevance" as the subset A and the event "a term occurs" as the subset B. To each event, a probability measure P assigns a real number in the unit range in the same way the coordination level assigns a weight to the document and the term.

An algebraic illustration of the relevance model is provided in the following. Suppose a universe of events (i.e., event space) Ω can be split into A and $\bar{A}$ such that

$$A \cup \bar{A} = \Omega \qquad A \cap \bar{A} = \emptyset$$

Similarly, the space is split according to the terms resulting from indexing the collection as follows:

$$B_i \cup \bar{B}_i = \Omega \qquad B_i \cap \bar{B}_i = \emptyset \qquad i = 1, \dots$$

Note that the same procedure can be applied for Boolean combinations of the terms.

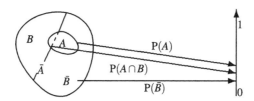

Fig. 1.18 Relevance model

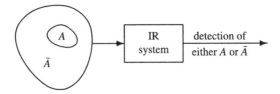

Fig. 1.19 Retrieval as a detection problem

1.4.1.1 Detection and Risk

The main problem with the relevance model is that A is not only unknown, it also changes for each information need. It follows that an IR system based on such a model has to detect whether a document is relevant although context continuously evolves and nothing is known if not for some content descriptors. As a system has to decide whether a document is relevant on the basis of a "signal" given by content descriptors, retrieval is viewed as a detection problem, as depicted in Fig. 1.19.

The dependency of A on the user's information need and ultimately on the context affecting the need is the reason why retrieval is affected by uncertainty, and therefore, the detection performed by the system is inevitably a statistical detection.

When performing such a detection, the system often (always) makes two errors: one error is to retrieve nonrelevant documents and the other error is to miss relevant documents. When making these errors, two costs arise: false alarm is the cost of retrieving (i.e., detecting) documents that are not relevant and loss of recall is the cost of not retrieving documents that are relevant. The detection costs are often encoded as follows:

- $c(A, \bar{A})$ is the cost of retrieving a document because it is decided that it is relevant (first A) when it actually is not relevant (second A); this is the cost of false alarm.
- $c(\bar{A}, A)$ is the cost of not retrieving a document because it is decided that it is not relevant when it actually is relevant; this is the cost of loss of recall.

It is defined $c(A, A)$ and $c(\bar{A}, \bar{A})$ although these costs are very often set to zero.

Although perfect retrieval, retrieval of all and only relevant documents, is impossible to obtain, optimal retrieval, retrieval of the largest number of relevant documents provided the maximum number of nonrelevant documents, can be obtained. To this end, risk is introduced. When the probability measure of the events and the costs is available, the risk of a detection can be computed for each event and can be defined as follows:

$$R(A|B) = c(A, A)\mathrm{P}(A \mid B) + c(A, \bar{A})\mathrm{P}(\bar{A} \mid B)$$

$$R(\bar{A}|B) = c(\bar{A}, A)\mathrm{P}(A \mid B) + c(\bar{A}, \bar{A})\mathrm{P}(\bar{A} \mid B)$$

An IR system decides to retrieve the documents in B when

$$R(A \mid B) < R(\bar{A} \mid B)$$

and this happens if and only if

$$P(A \mid B) > \frac{c(A, \bar{A}) - c(\bar{A}, \bar{A})}{c(\bar{A}, A) + c(A, \bar{A}) - c(A, A) - c(\bar{A}, \bar{A})}$$

The latter equation shows that the costs are like knobs operating on a device that emits documents. When the cost of loss of recall that can be accepted by the system or the end user increases, the number of retrieved documents increases since the threshold decreases; this can be explained by the fact that when $c(\bar{A}, A)$ increases, the miss of relevant documents is less tolerated, and therefore, the system accepts to retrieve further documents. How the costs of loss of recall and false alarm explain the threshold is depicted in Fig. 1.20 when the costs of correct decision are null.

1.4.1.2 Probability Ranking Principle

The effectiveness of the relevance model rests on the PRP:

> If a reference retrieval system's response to each request is a ranking of the documents in the collection in order of decreasing probability of relevance to the user who submitted the request, where the probabilities are estimated as accurately as possible on the basis of whatever data have been made available to the system for this purpose, the overall effectiveness of the system to its user will be the best that is obtainable on the basis of those data.

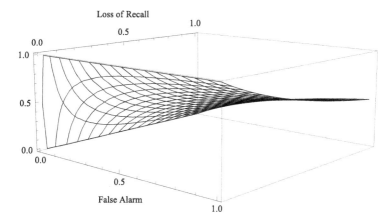

Fig. 1.20 Probability of relevance threshold as function of loss of recall and false alarm

The PRP was first introduced by Maron and Kuhns (1960), and a discussion is reported by Robertson (1977). The reason why the PRP is significant in IR is the link between the principled approach to ranking and the effectiveness measures. The risks defined within the PRP are actually probabilistic definitions of recall and fallout. Thanks to this probabilistic view, the maximization of the probability of relevance determines the maximization of recall at a given maximum tolerated fallout. In this way, the PRP states that the retrieval effectiveness is optimized when recall is maximum for each fixed cost of fallout. In practice, the principle tells to determine the Bs such that the fallout is less than a given threshold and to choose the B that maximizes the recall among the previously determined Bs.

1.4.1.3   Model Estimation

Corresponding to the events of the probability space, two main random variables can be defined:

$$X_j(\omega) = 1 \qquad\qquad t_j \text{ occurs in } \omega$$
$$X_A(\omega) = 1 \qquad\qquad \omega \text{ is relevant}$$

For example, suppose the following four documents $\omega_1 =$ "apple banana", $\omega_2 =$ "apple banana cherry", $\omega_3 =$ "banana apple", and $\omega_4 =$ "banana" are indexed and three terms are extracted from the documents. We have that

$$X_{\text{apple}}(\omega_1) = 1 \qquad X_{\text{banana}}(\omega_1) = 1 \qquad X_{\text{cherry}}(\omega_1) = 0$$
$$X_{\text{apple}}(\omega_2) = 1 \qquad X_{\text{banana}}(\omega_2) = 1 \qquad X_{\text{cherry}}(\omega_2) = 1$$
$$X_{\text{apple}}(\omega_3) = 1 \qquad X_{\text{banana}}(\omega_3) = 1 \qquad X_{\text{cherry}}(\omega_3) = 0$$
$$X_{\text{apple}}(\omega_4) = 0 \qquad X_{\text{banana}}(\omega_4) = 1 \qquad X_{\text{cherry}}(\omega_4) = 0$$

In general, when k terms are extracted, the multidimensional random variable

$$X = (X_1, \ldots, X_k)$$

can be defined, thus yielding up to 2^k possible outcomes. Let B be a subset of the set of elementary events mapped to a given outcome of X; for example, when $X = (0, 1, 0)$, we have that $B = \{\omega_2, \omega_4\}$. These subsets can be restricted to the subset A of relevant documents, thus obtaining the outcomes of X conditioned to relevance. When the elementary events ω are assigned a probability measure $P(\omega)$, the following probabilities can be computed:

$$P(B \mid A) \qquad P(B \mid \bar{A})$$

The actual use of the relevance model requires the estimation of $P(B|A)$ and $P(B|\bar{A})$. The documents in B are described by k properties which are in turn described by a random variable X. The simplest approach is binary:

$$X_j(\omega) = 1 \quad \text{term } j \text{ occurs in } \omega \quad \omega \in B$$

Suppose B can be mapped to X. It follows that

$$P(B|A) = P(X = x|X_A = 1) \qquad P(B|\bar{A}) = P(X = x|X_A = 0)$$

When $X = x$ is $X_1 = x_1, \ldots, X_k = x_k$,

$$P(X = x|X_A = 1) = P(X_1 = x_1, \ldots, X_k = x_k|X_A = 1)$$

Since there are 2^k possible outcomes, the number of estimations is exponential, and its estimation is in practice infeasible although k might be small; this problem is known as the curse of dimensionality and may be addressed by assuming conditional stochastic independence between the X_js defined as follows[2]:

$$P(X_1 = x_1, \ldots, X_k = x_k|X_A = 1) = \prod_{j=1}^{k} P(X_j = x_j|X_A = 1)$$

and

$$P(X_1 = x_1, \ldots, X_k = x_k|X_A = 0) = \prod_{j=1}^{k} P(X_j = x_j|X_A = 0)$$

Suppose

$$p_j = P(X_j = 1 \mid X_A = 1) \qquad q_j = P(X_j = 1 \mid X_A = 0)$$

It follows that

$$P(X = x \mid X_A = 1) = \prod_{j=1}^{k} p_j^{x_j}(1 - p_j)^{1-x_j}$$

$$P(X = x \mid X_A = 0) = \prod_{j=1}^{k} q_j^{x_j}(1 - q_j)^{1-x_j}$$

---

[2]Cooper (1995) showed that this assumption can be weakened.

The application to IR gives the likelihood ratio of the Binary Independence Retrieval (BIR) model where the likelihood ratio is

$$L(x) = \frac{P(X = x \mid X_A = 1)}{P(X = x \mid X_A = 0)} = \frac{\prod_{j=1}^{k} p_j^{x_j}(1 - p_j)^{1-x_j}}{\prod_{j=1}^{k} q_j^{x_j}(1 - q_j)^{1-x_j}}$$

and the log-likelihood ratio is

$$\ell(x) = \log L(x)$$

that is,

$$\ell(x) = \sum_{j=1}^{k} x_j w_j + \sum_{j=1}^{k} \log \frac{1 - p_j}{1 - q_j}$$

where $w_j = \log \frac{p_j(1-q_j)}{q_j(1-p_j)}$ is called Term Relevance Weight (TRW).

In the relevance model, the query is directly not modeled, whereas relevance is modeled since it is represented as a subset of documents. However, the BIR model requires query modeling due to efficiency reasons since the calculation of the log-likelihood would require k additions. To reduce the computational cost, which might be large when k is large, it is supposed that a query is given as input so that the summation is limited to the TRWs of the query terms.

The estimation of the TRWs is based on the maximum livelihood estimators (MLEs) of the p_js and q_js. Provided a training subset of documents, the following Maximum Likelihood Estimators (MLEs) are used:

$$\hat{p}_j = \frac{r_j + \frac{1}{2}}{R + 1} \qquad \hat{q}_j = \frac{n_j - r_j + \frac{1}{2}}{N - R + 1}$$

where R is the number of relevant documents in the training set, $r_j \leq R$ is the number of relevant documents indexed by term j, N is the number of documents, and n_j is the number of documents indexed by term j; the constants are commonly utilized to smooth the estimators.

1.4.1.4   Best Match N. 25

Robertson and Walker (1994) proposed a variation of the TRW which became one of the most effective weighting schemes. Best Match N. 25 (BM25) basically multiplies the TRW by a saturation component, thus obtaining the following weight:

$$w_{ij} = \text{TRW}_j \text{SATURATION}_{ij}$$

where the first component is the TRW also known as Robertson and Sparck Jones (1976)'s weighting scheme defined as

$$\text{TRW}_j = \log \frac{r_j + 0.5}{R - r_j + 0.5} - \log \frac{n_j - r_j + 0.5}{N - n_j - R + r_j + 0.5}$$

The TRW is multiplied by a saturation component

$$\text{SATURATION}_{ij} = \frac{(k_1 + 1)f_{ij}(k_3 + 1)g_j}{(k + f_{ij})(k_3 + g_j)}$$

of term j in document i. For each document, the saturation component is a monotonically increasing function of the frequency, f_{ij}, of j in i. The shape of this function is tuned by a number of parameters and variables; $k = k_1((1 - b) + b\frac{l_i}{l})$, l is the average document length, l_i is the length of document i, b is a parameter (usually 0.75), k_1 and k_3 are parameters (usually, 1.2 and something between 7 and 1000, respectively), and g_j is the frequency of term j in the query.

1.4.1.5 Relevance Feedback

RF in the relevance model consists of modifying the TRWs. The iterative process of RF begins with the situation in which no relevance data are available, that is, $R = 0$. It follows that at the beginning, ranking is computed by the following function:

$$g^{(0)}(z) = \sum_{j=1}^{k} z_j w_j^{(0)}$$

where

$$w_j^{(0)} = \log \frac{N - n_j + \frac{1}{2}}{n_j + \frac{1}{2}}$$

At step $t = 1, 2, \ldots$ of RF, the following function is used instead:

$$g^{(t)}(z) = \sum_{j=1}^{k} z_j w_j^{(t)}$$

where

$$w_j^{(t)} = \log \hat{p}_j^{(t)} + \log 1 - \hat{q}_j^{(t)} - \log \hat{q}_j^{(t)} - \log 1 - \hat{p}_j^{(t)}$$

$$\hat{p}_j^{(t)} = \frac{r_j^{(t)} + a^{(t)}}{R^{(t)} + b^{(t)}} \qquad \hat{q}_j^{(t)} = \frac{n_j - r_j^{(t)} + c^{(t)}}{N - R^{(t)} + d^{(t)}}$$

It is usually assumed that

$$a^{(t)} = c^{(t)} = \frac{1}{2} \qquad b^{(t)} = d^{(t)} = 1$$

but the theory of Bayesian statistics may provide additional hints.

1.4.2 Language Models

This model is named after the fact that it explicitly represents language, while
relevance is not explicitly represented; in the literature, both terms "language
model (LM)" and "language models (LMs)" are utilized depending on whether
one is referring to the class of probabilistic retrieval models sharing the properties
described in this section or to the specific probabilistic space describing how
language can statistically be described. In the following, the term is meant to
indicate the class of probabilistic retrieval models.

According to a LM, there is an author of a document thinking about the queries a
possible end user would formulate to retrieve the document; for example, an author
may write the sentences of the document in a way that they contain the answers to
the users' questions. In doing that, the author writes the document using queries and
variations of them, although he is assumed to have a good idea of the user's need.
On the other end, there is a user assumed to have a good idea of what he is searching
for.

So that documents and queries can be matched, the author and the user are
assumed to use an effective language and the same language. Most importantly, it is
also assumed that the documents generated by the authors are relevant to the user's
information need. The LM is indeed known as a generative model since it describes
how language is generated; in particular, the Language Model (LM) describes how
documents and queries are generated and how a query can be viewed as the outcome
of the generation fueled by a document. Figure 1.21 gives a pictorial description of

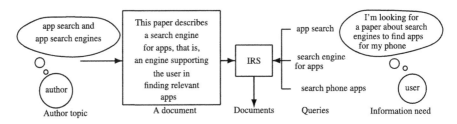

Fig. 1.21 Metaphor of the Language Model

how the LM is used in IR. On the left-hand side of the figure, an author is thinking about "app search and app search engines" and is writing a document pertinent to this topic. On the right-hand side of the figure, a user is thinking about his own information need and formulates a series of queries describing the aspects of the information need. At the center of the figure, an IR based on the LM matches the author's document and the user's queries and decides whether the document and the queries derive from the same language. The basic assumption of the LM is that a document contains information relevant to the user's information need when it is of the same language as the user's queries.

1.4.2.1 Documents and Queries as Languages

Algebraically, the LM consists of symbols, strings of symbols (i.e., n-grams), and a probability function defined on these strings. Let s be a symbol. A language is defined as a set of symbols:

$$\{s_1, \ldots, s_N\}$$

Given a language, an n-gram is a sequence of n symbols drawn from the language expressed as

$$s_{(1)} \ldots s_{(n)} \qquad n > 0$$

When $n = 1$, the n-gram is called unigram; if $n = 2$, it is called bigram; if $n = 3$, it is called trigram. A language can be viewed as an urn like the one in Fig. 1.22 from which symbols are drawn to form n-grams. Whenever a document or a query is formed, a series of symbols are drawn from the urn, and the outcome of this process is an abstract representation of the document or the query. Probabilistically, an n-gram is an experimental outcome; for example, suppose $L = \{s_1, s_2, s_3\}$ and $n = 2$. Sampling two symbols with replacement yields the following sampling space (i.e., the space of all the possible outcomes):

$$s_1 s_1, s_1 s_2, s_1 s_3, s_2 s_1, s_2 s_2, s_2 s_3, s_3 s_1, s_3 s_2, s_3 s_3$$

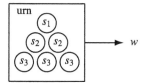

Fig. 1.22 Language as an urn

"With replacement" means that using the same symbol in an n-gram is possible; indeed, natural languages work with replacement. A sampling space contains N^n n-grams when order matters and contains $\binom{N}{n}$ n-grams when order does not matter; the space is therefore very large, thus explaining the high variety of a natural language in which the words are symbols, whereas documents and queries are n-grams.

The LM is a language provided with a probability function; for example, the language in Fig. 1.22 is an instance of the LM where $P(s_j) = j/6$. In IR, a language model is often built from a document or a group of documents. Consider the following document:

upon the bench the goat lives, under the bench the goat dies

After removing stop words and stemming words, the resulting language is

$$L = \{\text{bench, goat, live, die}\}$$

Suppose $n = 2$; the LM is given by a language and by the probability function

$$P(s_1, s_2) = P(s_1|s_2)P(s_2)$$

where

$$\{s_1, s_2\} \in \{\{\text{bench, goat}\}, \{\text{goat, live}\}, \{\text{live, bench}\}, \{\text{goat, die}\}\}$$

1.4.2.2 Query Language Model

Among the various instances of LM, the Query Language Model (QLM) is the one mostly used in IR. According to the QLM, documents are samples of a language and queries are LMs. The IR system designed according to the QLM looks for the most likely document given a query:

$$B^* = \arg_B \max P(B \mid Q)$$

where Q is a query event and B is a document event. Documents are ranked by $P(B \mid Q)$. However, Q is not completely known: the language is known but the probability is unknown. Therefore, the Bayes theorem is applied to obtain

$$P(B \mid Q) = \frac{P(Q \mid B)P(B)}{P(Q)}$$

thus swapping the roles played by query and document. Given a query, $P(Q)$ is constant and therefore, the ranking of the documents is not affected. $P(B)$ is assumed to be either uniform and then not affecting the document ranking or estimated by external sources such as PageRank or other query-independent measures of document qualities. With regard to probability estimation, as Q is regarded as an n-gram, we have

$$P(Q \mid B) = p_B(s_{(1)} \ldots s_{(n)}) = p_B(s_{(1)})p_B(s_{(2)}|s_{(1)}) \cdots p_B(s_{(n)} \mid s_{(n-1)} \cdots s_{(1)})$$

1.4.2.3 Mixture and Smoothing

Due to the curse of dimensionality encountered when the relevance model was described, stochastic independence has to be assumed, thus obtaining

$$P(Q|B) = p_B(s_{(1)}) \cdots p_B(s_{(n)})$$

where

$$p_B(s_{(j)}) = \frac{f(s_{(j)}, B)}{\sum_{j=1}^{n} f(s_{(j)}, B)}$$

and $f(s, B)$ is the frequency of s in B. The problem that f might be 0 is solved either by a mixture as follows:

$$\hat{p}_B(s_{(j)}) = (1 - \lambda)\frac{f(s_{(j)}, B)}{\sum_{j=1}^{n} f(s_{(j)}, B)} + \lambda\frac{f(s_{(j)}, L)}{\sum_{j=1}^{n} f(s_{(j)}, L)} \tag{1.3}$$

or by smoothing as follows:

$$\hat{p}_B(s_{(j)}) = \frac{f(s_{(j)}, B) + a}{\sum_{w \in B} f(s_{(j)}, B) + a + b} \tag{1.4}$$

Mixture and smoothing are depicted in Fig. 1.23; mixture (Fig. 1.23a) consists of repeatedly sampling from an urn chosen with a given probability, whereas smoothing (Fig. 1.23b) consists of virtually modifying the urn before sampling. With mixture, first an urn is drawn with a given probability of the urn, and then the symbols are drawn from the selected urn. After collecting a sufficiently large set of outcomes together, a histogram of the frequencies of the symbols can be drawn. With smoothing, the process is like injecting additional symbols into each urn to avoid that an urn does not contain a symbol. After injecting these additional symbols, it is possible to proceed as with a mixture.

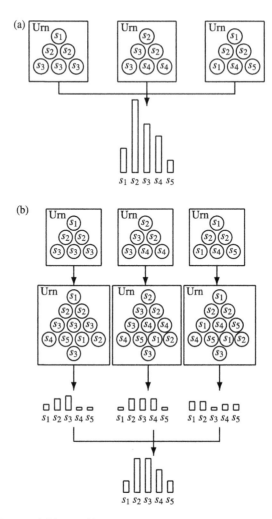

Fig. 1.23 (**a**) Mixture and (**b**) smoothing

1.5 Machine Learning

The use of Machine Learning (ML) models in IR may be explained by the difficulty in designing a single retrieval function that encompasses all the sources of evidence of a modern system (e.g., a search engine of the WWW or an "intelligent" system called to solve complex tasks). The difficulty in combining these sources is caused by the interaction between the user and system and the variety of context. ML provides an alternative approach to the problem of designing a system (i.e., a "machine") that can retrieve and rank documents in the best possible way for each

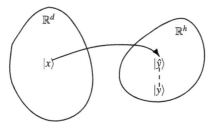

Fig. 1.24 Machine Learning as relation between spaces

query, and it gives the methods for learning a system to rank documents starting from training sets, i.e., from experience.

In ML, the reality of interest is viewed as a relation between points of the multidimensional real space. ML is concerned with the design and evaluation of algorithms and data structures for the estimation and prediction of relations between variables. In particular, ML aims at relating a point of one multidimensional space that represents the features of objects to a point of another multidimensional space that represents the patterns of the objects; for example, ML algorithms are designed to recognize handwritten text or elements depicted in images; to this end, the images are represented as feature vectors summarizing the properties of the images; a method (the machine) is then trained on training text manuscripts to relate each training vector to a label that summarizes a class of images; finally, it is evaluated on test data prepared to evaluate the effectiveness.

In mathematical terms, a method of ML assigns a point $|x\rangle \in \mathbb{R}^d$ to a point $|y\rangle \in \mathbb{R}^h$ as depicted in Fig. 1.24. The vector $|x\rangle$ is called "feature" (or feature vector) and $|y\rangle$ is called "pattern." While the feature vectors are in the multidimensional real space, the patterns may be qualitative labels (e.g., "high" or "low," "relevant" or "irrelevant") represented by basis vectors $|y\rangle \in \mathbb{R}^h$. During the training phase, pairs of feature vectors and patterns $(|x\rangle, |y\rangle)$ are observed and utilized to train a method; during the test phase, the trained method separates and assigns new feature vectors to the patterns with the aim of minimizing the risk of error.

1.5.1 Risk

In ML, we have the problem of finding an algorithm that relates some values (i.e., relevance labels or degrees) represented by $|y\rangle$ to some input values (i.e., documents) represented by $|x\rangle$. In general, there is a generator G of data $|x\rangle \in \mathbb{R}^d$ (i.e., a collection of documents) observed from an unknown probability distribution $p(x)$. There is also a supervisor S (i.e., a user of an IR system) who returns an output value represented by the $|y\rangle$s (i.e., an assessment about the relevance of a document) according to an unknown probability distribution $p(y|x)$. Therefore, the probability distribution $p(x, y) = p(y|x)p(x)$ is also unknown. There is finally an algorithm

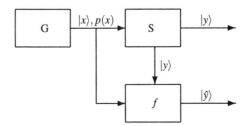

Fig. 1.25 Machine Learning setting

$f(|x\rangle, \alpha)$ depending on a set of parameters α and capable of associating an input value $|x\rangle$ to an output value $|\hat{y}\rangle$ which is often different from the point $|y\rangle$ that should be related to. This setting, which is inspired by Vapnik (1999), is depicted in Fig. 1.25.

In IR, f is indeed the retrieval function used to retrieve and rank documents in answer to the user's queries; for example, α represents a query, and an IR system has to learn α for retrieving and ranking documents to answer the query; alternatively, α may be the parameters of the Best Match N. 25 (BM25) weight function; in the event of learning to rank (Sect. 1.5.3), f is a method that reranks the documents retrieved by another retrieval function.

When applied to IR, ML aims at finding the best α where "best" means that a measure of utility is maximized or a measure of loss is minimized, these measures being related to the measures of effectiveness (e.g., Mean Average Precision (AP) (MAP)) utilized in IR. The common measure is based on risk, that is, the expected loss between the decision $f(|x\rangle, \alpha)$ and the true output value $|y\rangle$. Suppose loss is a function $L(|y\rangle, f(|x\rangle, \alpha))$ between the relevance label provided by a user and the label provided by f. Risk (or the risk function) is defined as

$$R(\alpha) = \sum_{|x\rangle \in \mathbb{R}^d, |y\rangle \in \mathbb{R}^h} L(|y\rangle, f(|x\rangle, \alpha))p(x, y)$$

The Empirical Risk Minimization (ERM) framework is adopted in ML to find the value of α which minimizes the risk function. The main problem of minimizing R is that $p(x, y)$ is unknown. However, a sample or training set

$$\{(|x_1\rangle, |y_1\rangle), \ldots, (|x_n\rangle, |y_n\rangle)\}$$

is available for estimating p. The ERM framework dictates that when the sample data are independently and identically distributed (i.i.d.), both R and the empirical risk function

$$R_n(\alpha) = \frac{1}{n} \sum_{i=1}^{n} L(|y_i\rangle, f(|x_i\rangle, \alpha))$$

converge to the same minimum, that is, when α_n minimizes R_n, it also minimizes R. Therefore, the empirical risk function can be utilized to find the desired minimum risk.

1.5.2 Separability

A significant pattern is the class. When the task of ML is classification, the input points have to be related to classes, and a class is represented by a point $|y\rangle$. Within classification, separability plays an important role. Suppose n points (i.e., feature vectors) are observed and placed in $\mathbb{R}^d$. Some of these points should belong to one class, whereas the other points should belong to other classes. Therefore, a method that separates the observed points into classes with the minimum risk of error is necessary; such a method is often called "classifier" and is mathematically represented by a subspace of $\mathbb{R}^d$ (also known as "iperplane").

The search for the subspace with the minimum risk of error should deal with two problems, that is, the points may be so close to each other that classification can be difficult if not even impossible, and they may also be collocated in a nonlinear way to make the task of a linear classifier (e.g., an iperplane) impossible. To illustrate separability and these two problems, consider three situations in which n points can be placed in $\mathbb{R}^d$. These points may be placed in $\mathbb{R}$ along a one-dimensional line (Fig. 1.26a), they may be placed in $\mathbb{R}^2$ in a bidimensional plane (Fig. 1.26b), or they may be placed in $\mathbb{R}^3$ in a tridimensional space (Fig. 1.26b). In each space of dimension d, a subspace of dimension $d - 1$ can be defined to separate the points in two subsets; these two subsets correspond to two patterns or to one pattern and the other patterns. The separating subspace is a point in a line, it is a line in a plane, or it is a plane in a space. The separating subspace is moved to a position to have the largest possible number of points of a pattern (e.g., black points) in a subset of points; for example, the separating point of Fig. 1.26a is placed to have the largest possible number of points of a pattern on one side and the points of the other pattern on the other side.

Separability is difficult because the points are very often scattered and the number of dimensions is often too small. While the various random directions along which the points are thrown cannot be controlled, the dimensionality of the space in which the points are placed can be varied. If an additional dimension were available, the separation of the set of points would be easier; for example, the points in the bidimensional space of Fig. 1.26b can be projected to the tridimensional space of Fig. 1.26c to allow us to separate them by a plane. The new dimension is added by a function ψ that projects one point of the d-dimensional space to the point of the $d + j$-dimensional space. In IR, the addition of a new dimension is often

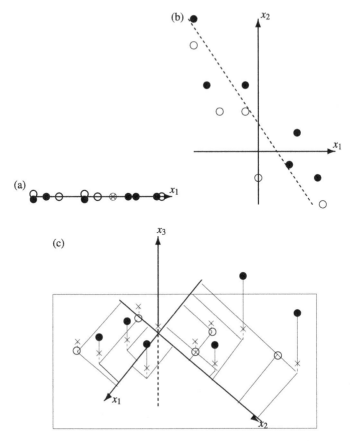

Fig. 1.26 Separability of a set of points. (**a**) Separating point. (**b**) Separating line. (**c**) Separating plane

implemented by the computation of weights; for example, if $\langle x| = (x_1, \ldots, x_d)$ is a vector of term frequencies in a document x, the additional dimensions may be IDFs, thus obtaining

$$
\psi(|x\rangle) = \begin{pmatrix} x_1 \\ \vdots \\ x_d \\ \log \frac{N}{x_1} \\ \vdots \\ \log \frac{N}{x_d} \end{pmatrix}
$$

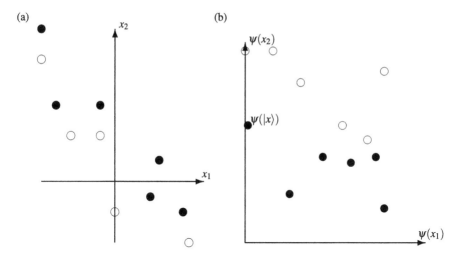

Fig. 1.27 Transforming a training set. (**a**) Not separable points. (**b**) Separable transformed points

After the transformation made by ψ, the training set becomes

$$\{(\psi(x_1), y_1), \ldots, (\psi(x_n), y_n)\}$$

The function ψ can also be utilized to better distribute the points in another space with the same dimensionality; for example, the training set depicted in Fig. 1.27a can be transformed by the arcsin function which transforms each coordinate in its arcsin to obtain the training set of Fig. 1.27b, which can be separated by a line more easily.

1.5.3 Learning to Rank

When applied to IR, the formulation of an ML model is in practice more elaborate. The basic idea is that an IR system that utilizes an ML model has to learn to rank a set of m documents to answer a representation of the user's information need (e.g., a query) given as input. The overall framework of learning to rank is depicted in Fig. 1.28 which is derived from Liu (2011)'s book. Suppose n queries $q_1, \ldots, q_n$ are given for training purposes. For each query q_i, an IR system retrieves a ranked list of $m^{(i)}$ documents. These n ranked lists are used by the ML system to learn the best parameters and eventually the best function f which is passed to the reranking system. The patterns represented by the points (vectors) $|y\rangle$ can have different forms depending on whether f has to relate an input point to a relevance degree, an order pair, or a ranked list; the details are addressed by Liu (2011).

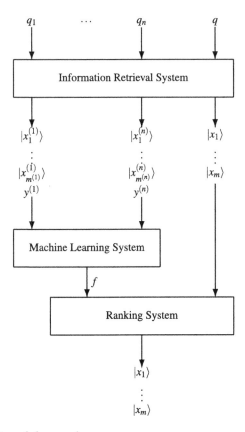

Fig. 1.28 Learning-to-rank framework

1.6   Suggested Readings

Mooers (1950) coined the term IR. Some years later, the works by Luhn (1958) and Luhn (1960) on automatic abstracting were published. Maron and Kuhns (1960) wrote one of the early papers on probabilistic IR. Later, Cleverdon and Mills (1963) reported on the experiments on automatic indexing. In the same year, Salton (1963) wrote a paper describing the statistical methods to measure semantic relationships between words such as synonymy and polysemy and to build networks of terms and documents. In 1966, Cleverdon et al. introduced the modern evaluation methodology based on test collections. Lovins (1968) described one of the early stemming algorithms, while Salton (1968) published his first textbook.

The use of statistical methods was established in the 1970s. In particular, Sparck Jones (1971) and Jardine and van Rijsbergen (1971) published the first results on automatic text classification, while Salton (1971) published the results of the first IR system based on vector spaces. In the same volume, Rocchio (1971) illustrated the RF method based on vector spaces. Sparck Jones and van Rijsbergen (1976) described the current evaluation methodology. Meanwhile, other textbooks were published by Salton and McGill (1983), Salton (1989), and van Rijsbergen (1979).

Thanks to the advent of the WWW, up-to-date textbooks were published in the 1990s and 2000s such as the books of Sparck Jones and Willett (1997), Manning and Schütze (1999), Frakes and Baeza-Yates (1992), Manning et al. (2008), and Croft et al. (2009) as for the algorithmic and modeling aspects. By contrast, the issues of user interaction and cognition were addressed by Ingwersen (1992) and Ingwersen and Järvelin (2005).

Boolean logic was of course introduced by Boole (1854), while one analysis of this logic in IR was provided by Cooper (1988). Nonclassical variants were presented in van Rijsbergen (1986), for example. The first elements of what was later named VSM were reported by Salton (1968) and later in Salton et al. (1975), while Salton (1979) reported on some mathematical aspects. Wong and Raghavan (1984) revisited the VSM, while Dubin (2004) revisited Salton's contribution to the field.

The basics of classical probability were established by Kolmogorov (1956). Probabilistic IR was introduced by Maron and Kuhns (1960) and Robertson and Sparck Jones (1976). The PRP was introduced by Maron and Kuhns (1960) and further investigated by Robertson (1977). The concept of TRW was addressed by Croft and Harper (1979), Robertson and Walker (1994), and Sparck Jones and van Rijsbergen (1976). Cooper (1995) relaxed some constraints on probability estimation. BM25 was proposed by Robertson and Walker (1994) and was later explained by Robertson and Zaragoza (2009).

The first approach to probabilistic IR based on LMs was proposed by Ponte and Croft (1998) and then by Zhai and Lafferty (2001) and Lavrenko and Croft (2001). Two surveys of this approach to probabilistic IR were published by Croft and Lafferty (2002), Lafferty and Zhai (2002), and Zhai (2008).

The classical reference of ML is by Vapnik (1999); that book is the condensed version of the book written by Vapnik in 1998. A survey of the methods to learn a system to rank documents was written by Liu (2011).

Chapter 2
Elements of Quantum Mechanics

After introducing the basic concepts of information retrieval in Chap. 1, this chapter briefly explains the main concepts of the quantum mechanical framework. In the introduction, we already noted that we selected the main concepts that may be linked to information retrieval. We first introduce observables and superposition; the former is usually known as random variable, while the latter is unknown in information retrieval. Probability has been introduced after superposition because quantum probability can be viewed as a generalization of probability, and this generalization is due to superposition. Interference and entanglement are two core concepts of the quantum mechanical framework and do not have any counterpart in information retrieval. Besides the core concepts of the quantum mechanical framework, we introduce detection since it is naturally linked to document retrieval and ranking. Finally, the chapter suggests some further readings.

2.1 Introduction

QM deals with the mathematical description of the motion and interaction of sub-atomic particles. QM established the impossibility of measuring physical systems at the microscopic level with arbitrary precision, thus legitimizing the axiom that physical systems at the microscopic level cannot be precisely and exhaustively observed using any device and that any observation can always be subjected to a probability measure of the degree to which the observed value is real. This chapter briefly explains the main concepts of the quantum mechanical framework. We selected the concepts of the quantum mechanical framework that may be linked to IR; an exhaustive survey of the subject would be infeasible since the literature is immense and covers a very complex network of topics.

A macroscopic object such as a ball obeys the laws of classical physics, that is, the physics before the advent of QM. In classical physics, the knowledge of the

© Springer-Verlag Berlin Heidelberg 2015
M. Melucci, *Introduction to Information Retrieval and Quantum Mechanics*,
The Information Retrieval Series 35, DOI 10.1007/978-3-662-48313-8_2

current state (e.g., the position) of an object (e.g., a ball) is sufficient to predict the future state of the object when the object is subjected to forces given that the initial conditions are perfectly known.

When many copies of the ball are put in an urn, one obtains a more complex object consisting of the container and of the copies of the ball. The urn still obeys the laws of physics; these laws are more complex since the urn is more complex than the individual balls. Although these laws are more complex than those governing an individual ball, it is still possible to apply the principles of classical physics and predict the future of the urn.

A macroscopic object such as a ball consists of a huge number of microscopic objects, i.e., atoms. Nineteenth-century scientists found that the law of physics applied to macroscopic objects such as balls and skyscrapers could not be applied to the microscopic world made of atoms and photons. What was found is that the law of motion, for instance, that characterizes the ball cannot be applied to each individual atom, nor can these laws be derived from other laws of physics in the same way as the classical laws can be applied to the urn which contains the balls.

Another discovery stemming from the physics of the last century has been that the measurement of microscopic objects is far different from the measurement of macroscopic objects. When measuring the speed of a moving ball, one puts a sensor at the beginning of a path, puts another sensor at the end of the path, and counts the number of time units necessary for the ball to move along the path. The apparatus of measurement consisting of sensors, path, and clock does not disturb the movement of the ball; actually, the forces, e.g., gravity generated from the apparatus, are not strong enough to disturb the movement of the ball.

In contrast, when measuring the speed of a moving microscopic object like a photon, one might put a sensor at the beginning of a path and another sensor at the end of the path and then count the number of time units necessary for the photon to move along the path. The apparatus of measurement disturbs the movement of the photon because the particles of the (macroscopic) measurement apparatus are of the same kind as the microscopic object, and the energy produced by the sensors is very strong relative to the energy produced by the moving photon.

Something similar to the measurement of the speed of a photon can be imagined in the macroscopic world. Consider, for instance, the measurement of the exact amount of asbestos of a skyscraper. Since searching for every fiber of asbestos would be very expensive and tedious, a demolition apparatus and a device filtering out the fibers might be utilized (see also the description of Meglicki (2008)). However, the demolition of the skyscraper would produce an amount of pulverized debris which would give us only an approximation of the exact amount of the asbestos and of other chemical elements in the original building. If the skyscraper could be replicated in exactly the same way and the demolition could be repeated in exactly the same way, it would be possible to calculate some distributions of the frequencies of the measured quantities of asbestos associated with these experiments, and an expected quantity of asbestos might be estimated. However, the measurement of other properties would become impossible since the demolition of the skyscraper destroys many other properties that would be observable when the skyscraper was intact, for example, height.

The measurement of microscopic objects is similar to the measurement of the volume of asbestos in the demolished skyscraper. Even if it were possible, the repetition of the measurement in exactly the same way would be very difficult and prone to error. In contrast, the measurement in classical physics does not destroy the object of measurement, and the uncertainty of measurement is only due to the possibly large number of objects and to the errors of measurement which can be bound within predefined limits by repeating the measurement many times.

This is the essence of uncertainty in QM: the empirical evidence that can be collected to estimate a quantity (e.g., of asbestos) is about one individual (e.g., skyscraper) having a particular quantity and not about an ensemble of individuals having that quantity. As measuring a property would destroy or at least alter the object under measurement, any other property that interferes (i.e., does not commute) with it could not be measured, and thus counting the number of individuals having a given quantity is simply not possible or does not make any sense. The latter consequence is known as incompatibility between two observables that cannot commute; the measurement of property A interferes with the measurement of property B, and the outcome of the measurement of B would have been different were it obtained before the outcome of the measurement of A.

The chapter is organized in the following sections. Section 2.2 is devoted to observables; this term is rather a noun than an adjective for drawing special attention to the devices and to the corresponding mathematical objects used to measure physical systems. Section 2.3 is devoted to a special feature of QM, i.e., superposition; it refers to the states of a system and to the possibility that a system can "at the same time" be in many different states; this is one of the least understood phenomena of physics. Section 2.4 introduces probability in the quantum mechanical framework; quantum probability departs from the classical probability because of superposition. Section 2.5 illustrates interference, which is another consequence of superposition and causes the invalidity of the laws of classical probability in the quantum mechanical framework. Section 2.6 is devoted to another mysterious phenomenon studied by QM, i.e., entanglement; it is built when a pair of photons is generated in a way to constrain the description of one photon dependently on the description of the other photon. Section 2.7 situates signal detection in the quantum mechanical framework. Section 2.8 concludes the chapter with some suggested readings.

2.2 Observables

The real world in which everybody lives can usually be observed, and what can be known to humans must be observed. Observation produces data, thus providing a representation of objects such that whatever the object's size, the degree of

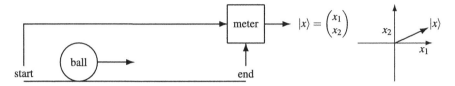

Fig. 2.1 State vector in the real bidimensional space

precision, and completeness of what we know about the real-world objects can be basically only given by the observed data.

The data resulting from an observation may be qualitative such as assessments about size (e.g., the tags "large" or "tall" given to a tree) or ability to support somebody in performing a task (e.g., the tag "relevant" given to a document when retrieved to meet an information need). Alternatively, the data used to represent the real world are quantitative, such as frequency (e.g., the number of occurrences of a term in a textual document or the number of red pixels in an image document) or polarization (e.g., the direction of oscillation of light photons).

The data are variables since they vary according to the observed objects; in statistics, the data observed in the real-world objects are modeled by random variables since the way the data vary can be described by some random law or probability measure. How these data vary and how they are correlated provide information about the real world.

The data observed in the real-world objects are in summary an abstraction of the objects, and the quantitative nature of the data makes the calculation of further data, and in this way the extrapolation of further information about the objects, possible.

In classical physics, the set of data collected during the observation of an object is called "state," and it is usually represented as a vector; if d distinct data are observed, the state vector has dimension d. The state vector is defined over the real field since the physical properties measured in an object need to correspond to the human perception of these properties; for example, the velocity of a ball is a real number since it is the ratio between two physical properties (Fig. 2.1).

In QM, the state vector is defined in the complex field because the solution of the equations that form the models explaining the microscopic world often can only be found in the complex field. Moreover, a state is not only a vector; it is on the contrary an operator often represented by a matrix over the complex field and used to compute probabilities. In QM, the variables used to describe real-world objects are called observables, thus emphasizing the fact that the information that can be achieved from a microscopic object can only be the outcome of an observation. Therefore, a real-world object can be observed using one or more observables. Each observable produces some data which are collected together with the data produced by other observables to provide a representation of the object; for example, the photon of a light beam can be described by an observable that yields the orientation of the photon with respect to a polarization.

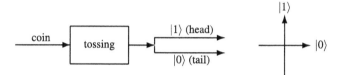

Fig. 2.2 Coin tossing in the vector space

2.2.1   Vector Space Basis and Projector

In QM, an observable is mathematically represented by the basis of a vector space over the multidimensional complex field. The dimensionality d of the vector space is the number of distinct values taken by the observable; for example, the observable that describes the outcome of coin tossing takes $d = 2$ values (i.e., head or 1 and tail or 0) and the vector space, and then the basis is bidimensional. It follows that the number of basis vectors is the number of distinct values taken by the observable; for example, coin tossing corresponds to a basis of two vectors as depicted in Fig. 2.2 and formalized as follows:

$$
\begin{array}{ccc}
\text{Outcome} & \text{Value} & \text{Ket} \\
\text{head} & 1 & |1\rangle = \begin{pmatrix} 1 \\ 0 \end{pmatrix} \\[2ex]
\text{tail} & 0 & |0\rangle = \begin{pmatrix} 0 \\ 1 \end{pmatrix}
\end{array}
$$

The second column of the table includes the values taken by a variable used in coin tossing to represent the possible outcomes. These values are often reused to label the kets, e.g., $|0\rangle$, yet other symbols may be used.

Note that the basis vectors of an observable are mutually orthogonal, thus meaning that the events (e.g., "head" and "tail") are mutually exclusive; for example, two orthogonal basis vectors corresponding to two mutually exclusive events may be

$$
|1\rangle = \begin{pmatrix} 1 \\ 0 \end{pmatrix} \qquad |0\rangle = \begin{pmatrix} 0 \\ 1 \end{pmatrix}
$$

but the following orthogonal vectors

$$
\begin{pmatrix} \frac{1}{\sqrt{2}} \\ \frac{1}{\sqrt{2}} \end{pmatrix} \qquad \begin{pmatrix} \frac{1}{\sqrt{2}} \\ -\frac{1}{\sqrt{2}} \end{pmatrix}
$$

are also representing mutually exclusive events.

An alternative way to describe observables is using projectors by leveraging the connections with the Boolean logic introduced in Sect. 1.2. To explain the relationship between observables and projectors and why projectors may eventually be preferred to vectors, consider the vector space basis of Fig. 2.2. A basis vector represents a proposition such as "head," whereas the other basis vector represents the opposite proposition such as "tail."

Given a basis vector, a mathematical function is needed to assign the truth value to the proposition represented by a basis vector. Consider the event "head" represented by the monodimensional vector space (i.e., a line) spanned by $|1\rangle$; this vector spans all the vectors of this space. This monodimensional vector space and the corresponding vector $|1\rangle$ answer the question about coin tossing. The same vector can be used to ask the same question; it is expected that the answer is 1 if and only if a head was observed, 0 otherwise.

The only function that can provide these answers is the inner product since the basis vectors are unit and mutually orthogonal vectors, that is, $\langle 1|1\rangle = 1$ and $\langle 1|0\rangle = 0$. If a vector were not a unit and on the contrary was $c|1\rangle = |c\rangle$, the inner product would yield many results such as the number of distinct values of c.

The proposition opposite to "head" is not necessarily represented only by $|0\rangle$ since $\langle 1|0\rangle = \langle 1|c0\rangle = c\langle 1|0\rangle = c0 = 0$ for all c. It can instead be represented by the subspace spanned by $|0\rangle$, which is indeed the set of vectors $c|0\rangle$ indexed by c,[1] or in general, it is represented by the subspace orthogonal to the subspace spanned by $|1\rangle$; if the space is bidimensional, the orthogonal subspace is spanned by $|0\rangle$, and it is then monodimensional; otherwise, the subspace orthogonal to the subspace spanned by $|1\rangle$ has $d - 1$ dimensions.

Therefore, a proposition is represented by a subspace and not by a vector. Suppose there are three possible observable values, each value corresponding to a basis vector as follows:

$$\text{win} \qquad |2\rangle \qquad = \begin{pmatrix} 1 \\ 0 \\ 0 \end{pmatrix}$$

$$\text{tie} \qquad |1\rangle \qquad = \begin{pmatrix} 0 \\ 1 \\ 0 \end{pmatrix}$$

$$\text{loss} \qquad |0\rangle \qquad = \begin{pmatrix} 0 \\ 0 \\ 1 \end{pmatrix}$$

The proposition opposite to "win" is of course "tie or loss"; therefore, the subspace including all the vectors of the matches that ended with ties or losses (i.e., $|0\rangle$ or $|1\rangle$)

---

[1] Recall that $|0\rangle \neq 0$ since it is a vector and not a scalar.

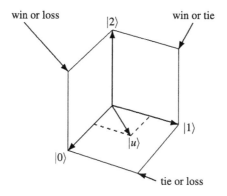

Fig. 2.3 Observable in the tridimensional space

is necessarily based on both $|0\rangle$ and $|1\rangle$, thus making it a bidimensional space (i.e., a plane); indeed, the plane spanned by $|0\rangle$ and $|1\rangle$ is orthogonal to the line spanned by $|2\rangle$. The latter fact can be viewed in Fig. 2.3.

If an observable has two values as depicted in Fig. 2.2, there are two projectors (i.e., two basis vectors) which have to be mutually orthogonal since each observable value excludes any other value. The projectors can be defined as follows:

$$|0\rangle\langle 0| = \begin{pmatrix} 0 & 0 \\ 0 & 1 \end{pmatrix} \qquad |1\rangle\langle 1| = \begin{pmatrix} 1 & 0 \\ 0 & 0 \end{pmatrix}$$

and it can be checked that these projectors are mutually orthogonal, that is,

$$|i\rangle\langle i|j\rangle\langle j| = \langle i|j\rangle|i\rangle\langle j| = 0 \qquad i \neq j$$

since $\langle i|j\rangle = 0$ for all $i \neq j$.

Adding $|0\rangle\langle 0|$ and $|1\rangle\langle 1|$ is expressing that the observable has value 0 or value 1, that is,

$$\mathbf{1} = |0\rangle\langle 0| + |1\rangle\langle 1|$$

while

$$\mathbf{1} - |1\rangle\langle 1|$$

corresponds to observing all the observable values except 1.

In general, adding projectors corresponds to disjoining propositions as illustrated in Sect. 1.2; for example, if an observable has three values as depicted in Fig. 2.3, there are three projectors (i.e., three basis vectors) which have to be mutually orthogonal since each observable value excludes any other value. The projectors can be defined as follows:

$$|0\rangle\langle0| = \begin{pmatrix} 0 & 0 & 0 \\ 0 & 0 & 0 \\ 0 & 0 & 1 \end{pmatrix} \qquad |1\rangle\langle1| = \begin{pmatrix} 0 & 0 & 0 \\ 0 & 1 & 0 \\ 0 & 0 & 0 \end{pmatrix} \qquad |2\rangle\langle2| = \begin{pmatrix} 1 & 0 & 0 \\ 0 & 0 & 0 \\ 0 & 0 & 0 \end{pmatrix}$$

Similarly to the case of two projectors, it can be checked that these projectors are mutually orthogonal, that is,

$$|i\rangle\langle i|j\rangle\langle j| = \langle i|j\rangle|i\rangle\langle j| = \mathbf{0} \qquad i \neq j \quad i = 0,1,2 \quad j = 0,1,2$$

since $\langle i|j\rangle = 0$ for all $i \neq j$. Adding $|0\rangle\langle0|$ and $|1\rangle\langle1|$ is expressing that the observable has value 0 or value 1, while

$$\mathbf{1} - |1\rangle\langle1|$$

corresponds to observing 0 or 2.

Projectors are used to check whether the propositions they implement are true or false. To this end, the product between a projector and a vector that represents an element is calculated. Consider a projector $\mathbf{P}$ and a vector $|v\rangle$ representing an element which can make the proposition either true or false. The proposition implemented by $\mathbf{P}$ is true if

$$\mathbf{P}|v\rangle = |v\rangle$$

that is, if $|v\rangle$ belongs to the subspace represented by the projector; for example, suppose $|v\rangle$ represents "head," then we have that

$$\begin{pmatrix} 1 & 0 \\ 0 & 0 \end{pmatrix} |1\rangle = |1\rangle\langle1|1\rangle = |1\rangle 1 = |1\rangle \qquad \begin{pmatrix} 0 & 0 \\ 0 & 1 \end{pmatrix} |1\rangle = |0\rangle 0 = \mathbf{0}$$

and that

$$\langle1| \begin{pmatrix} 1 & 0 \\ 0 & 0 \end{pmatrix} |1\rangle = 1 \qquad \langle1| \begin{pmatrix} 0 & 0 \\ 0 & 1 \end{pmatrix} |1\rangle = 0$$

where 1 means "true" and 0 means "false." It can be checked that

$$\langle v|\mathbf{1}|v\rangle = \langle v|v\rangle = 1 \quad \text{for all } |v\rangle$$

However, note that a vector $|u\rangle$ is contained by the plane spanned by $|0\rangle$ and $|1\rangle$ of Fig. 2.3. It follows that

$$\langle u|(|0\rangle\langle0| + |1\rangle\langle1|)|u\rangle = \langle u|0\rangle\langle0|u\rangle + \langle u|1\rangle\langle1|u\rangle = |\langle u|0\rangle|^2 + |\langle u|1\rangle|^2 = 1$$

that is, $|u\rangle$ makes the proposition represented by $\mathbf{1}$ true, but it does not make the individual propositions represented by $|0\rangle$ and $|1\rangle$ true. Therefore, the subspace spanned by $|0\rangle$ and $|1\rangle$ contains infinitely many more vectors which make the proposition implemented by the projector true than the individual subspace spanned by $|0\rangle$ and the individual subspace spanned by $|1\rangle$. This situation is the same case observed in Sect. 1.2.4.5 and will be addressed in Sect. 2.2.2.

The extension of the algebraic way based on projectors to represent the events and propositions defined from one observable to the case of two or more observables is quite straightforward; however, the pictorial illustration is impossible since the number of dimensions would be greater than three.

Consider a pair of observables. Suppose that the values of both observables are $\{0, 1\}$. When an observable is considered in combination (or product) with the other, there are four possibilities: 00, 01, 10, and 11 where the first value refers to the first observable and the second value refers to the second observable. When a combination (or product) of three observables is considered, there are eight possibilities: 000, 001, 010, 011, 100, 101, 110, and 111 where the first value refers to the first observable, the second value refers to the second observable, and the third value refers to the third one. In general, the product of n binary observables is an observable whose values are in a space of dimensionality 2^n; when n is quite large, the string of bits is replaced by the corresponding number expressed in the decimal base (e.g., $0, 1, 2, 3, 4, 5, 6, 7$ when $n = 3$); for example, if tossing two coins is considered, the basis vector corresponding to the event that two tails occur is

$$|00\rangle = \begin{pmatrix} 0 \\ 0 \\ 0 \\ 1 \end{pmatrix}$$

This vector is obtained by the tensor product of the basis vectors corresponding to the single events of each observable; for example, tail is expressed by the basis vector

$$|0\rangle = \begin{pmatrix} 0 \\ 1 \end{pmatrix}$$

for both coins, and the tensor product of these two basis vectors is

$$|0\rangle \otimes |0\rangle = \begin{pmatrix} 0|0\rangle \\ 1|0\rangle \end{pmatrix} = \begin{pmatrix} 0 \begin{pmatrix} 0 \\ 1 \end{pmatrix} \\ 1 \begin{pmatrix} 0 \\ 1 \end{pmatrix} \end{pmatrix} = \begin{pmatrix} 0 \times 0 \\ 0 \times 1 \\ 1 \times 0 \\ 1 \times 1 \end{pmatrix} = \begin{pmatrix} 0 \\ 0 \\ 0 \\ 1 \end{pmatrix}$$

Using the tensor product, the basis vectors of the observable values $01, 10, and 11$ are, respectively,

$$|01\rangle = \begin{pmatrix} 0 \\ 0 \\ 1 \\ 0 \end{pmatrix} \qquad |10\rangle = \begin{pmatrix} 0 \\ 1 \\ 0 \\ 0 \end{pmatrix} \qquad |11\rangle = \begin{pmatrix} 1 \\ 0 \\ 0 \\ 0 \end{pmatrix}$$

Using the products $|ij\rangle\langle ij|$, the projectors are

$$|00\rangle\langle 00| = \begin{pmatrix} 0\,0\,0\,0 \\ 0\,0\,0\,0 \\ 0\,0\,0\,0 \\ 0\,0\,0\,1 \end{pmatrix} \qquad |01\rangle\langle 01| = \begin{pmatrix} 0\,0\,0\,0 \\ 0\,0\,0\,0 \\ 0\,0\,1\,0 \\ 0\,0\,0\,0 \end{pmatrix}$$

$$|10\rangle\langle 10| = \begin{pmatrix} 0\,0\,0\,0 \\ 0\,1\,0\,0 \\ 0\,0\,0\,0 \\ 0\,0\,0\,0 \end{pmatrix} \qquad |11\rangle\langle 11| = \begin{pmatrix} 1\,0\,0\,0 \\ 0\,0\,0\,0 \\ 0\,0\,0\,0 \\ 0\,0\,0\,0 \end{pmatrix}$$

It is then possible to compute the projector $|1?\rangle\langle 1?|$ of the event that head is tossed from the first coin independently of the second coin. The first head is tossed when either it is tossed and the second one is not or it is tossed and the second one is also tossed. Using projectors, we have that

$$|1?\rangle\langle 1?| = |10\rangle\langle 10| + |11\rangle\langle 11| = \begin{pmatrix} 1\,0\,0\,0 \\ 0\,1\,0\,0 \\ 0\,0\,0\,0 \\ 0\,0\,0\,0 \end{pmatrix}$$

2.2.2 Incompatibility

Two measurements are compatible when they are able to occur together without problems nor conflicts. In contrast, two measurements are incompatible when they cannot occur together without disturbing each other. Incompatibility is difficult to visualize, but an example from music may help; it is impossible to simultaneously come to knowledge of the instant of time and of the height of a note. This happens since the calculation of the height of a note needs the frequency analysis of sound, and this frequency can be estimated only if the sound is processed for a long enough time. Therefore, the height of a note results from a time interval which cannot correspond to a single instant of time.

In physics, the measurement of a quantity of a microscopic particle (e.g., photon) may be incompatible with the measurement of another quantity. If the position is measured first and the quantity of motion[2] is measured second, the state of the particle is different than the state obtained if the quantity of motion is measured first and the position is measured second. This lack of "commutativity" is called incompatibility. According to most interpretations of QM, incompatibility is caused by collapse. After the measurement of an observable, a particle is said to be collapsed when its state coincides with the observed value of the observable.

In contrast, classical physics lies in a compatible world. Suppose, for example, the velocity and position of a particle have to be measured; such a particle might be very large (e.g., the Earth) or very small (e.g., a grain of sand). Whatever the size, a particle can be idealized as a point in space; for example, although the Earth is obviously not a point, the size of the Earth can be ignored when calculating the position and the velocity of the planet around the Sun. Velocity and position are tied together by a system of equations relating the three spatial coordinates with time; in particular, velocity is the first derivative of position with respect to the time, and acceleration is the first derivative of velocity with respect to the time. These equations are exact in the sense that they provide exact values of velocity and acceleration provided position and time.

Consider two projectors **A** and **B**. When the commutativity between **A** and **B** holds, the observables of these projectors are called compatible. The cases considered in Sect. 1.2 assumed the commutativity between the projectors and then the compatibility between the observables; this is the foundation of the Boolean logic adopted in IR that makes the conjoint observation of, say, relevance and term occurrence possible.

Incompatibility is a key feature of the logic induced by the quantum mechanical framework (i.e., the so-called quantum logic), while the classical logic always has compatible propositions due to the commutativity of the conjunction operator; for example, when

$$\mathbf{A} = \begin{pmatrix} 1 & 0 \\ 0 & 0 \end{pmatrix} \qquad \mathbf{B} = \frac{1}{2} \begin{pmatrix} 1 & 1 \\ 1 & 1 \end{pmatrix}$$

compatibility is violated since

$$\mathbf{AB} \neq \mathbf{BA}$$

This example is depicted in Fig. 2.4, thus showing that the final result depends on the order of application of the projectors.

---

[2]This is the product of the velocity times the mass, and it is called momentum.

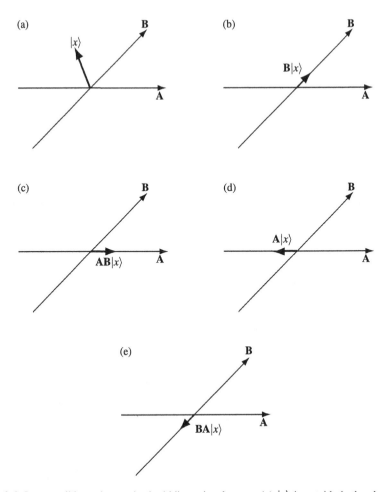

Fig. 2.4 Incompatible projectors in the bidimensional space. (**a**) $|x\rangle$ is outside both subspaces. (**b**) $|x\rangle$ is projected on **B**. (**c**) $\mathbf{B}|x\rangle$ is projected on **A**. (**d**) $|x\rangle$ is projected on **A**. (**e**) $\mathbf{A}|x\rangle$ is projected on **B**

Incompatibility causes the violation of the distributive law which is in contrast validated when the projectors are compatible, that is, when the classical Boolean logic is used; an example is depicted in Fig. 2.5. The figure shows a tridimensional vector space spanned by $|0\rangle, |1\rangle$, and $|2\rangle$; each of these vectors $|j\rangle$ originates a projector $|j\rangle\langle j|$ and spans a subspace $L(j)$. In that space, the projector $|\eta_0\rangle\langle\eta_0|$ corresponding to the monodimensional subspace $L(\eta_0)$ is spanned by $|\eta_0\rangle$, and the projector $|\eta_0\rangle\langle\eta_0| + |\eta_1\rangle\langle\eta_1|$ corresponds to a bidimensional subspace $L(\eta_0, \eta_1)$ spanned by $|\eta_0\rangle$ and $|\eta_1\rangle$. Note that the subspace $L(\eta_0, \eta_1)$ is also spanned by $|0\rangle$ and $|1\rangle$, that is,

$$L(\eta_0, \eta_1) = L(0, 1)$$

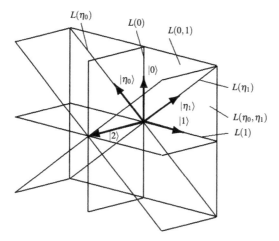

Fig. 2.5 Incompatible projectors in the tridimensional space

Following the explanation of Hughes (1989), consider the subspace

$$L(1) \wedge (L(\eta_0) \vee L(\eta_1))$$

provided that $\wedge$ means "intersection" and $\vee$ means "union." As

$$L(\eta_0) \vee L(\eta_1) = L(0, 1)$$

we have that

$$L(1) \wedge (L(\eta_0) \vee L(\eta_1)) = L(1) \wedge L(0, 1) = L(1)$$

However,

$$(L(1) \wedge L(\eta_1)) \vee (L(1) \wedge L(\eta_0)) = \emptyset$$

because

$$L(1) \wedge L(\eta_1) = \emptyset \qquad L(1) \wedge L(\eta_0) = \emptyset$$

therefore,

$$L(1) = L(1) \wedge (L(\eta_0) \vee L(\eta_1)) \neq (L(1) \wedge L(\eta_0)) \vee (L(1) \wedge L(\eta_1)) = \emptyset$$

thus meaning that the distributive law does not hold; hence, set operations cannot be applied to subspaces; the reader may want to look at (van Rijsbergen, 2004, pages 38–39) for a similar example.

2.3 Superposition

We introduce superposition by resorting to the notion of bit and then to the notion of qubit, which is the generalization of the bit in the quantum mechanical framework. Resorting to the qubit can be explained by the fact that it is the simplest form of quantum mechanical device, it can be directly implemented using atomic or subatomic particles, and it is essential in quantum computing, the latter being a "generalization" of computer science, IR included.

2.3.1 Bits

Flip-flop is one of the simplest devices used to implement bits. The description of the architecture of a flip-flop is outside the scope of this section; however, we would like to stress one characteristic of a flip-flop: the presence or the absence of an electric charge on the flip-flop determines the value stored in it. If there is no electric charge on the flip-flop, the value resulting from reading the device is zero. If there is an electric charge on the flip-flop, the value resulting from reading the device is one. This mechanism implies that a flip-flop has to store the value, that is, to keep the electric charge without destroying it over time or changing it through reading operations (Meglicki, 2008).

Algebraically, bits can be described by vectors, that is, the state of a bit is a vector. Recall that "bit" refers to two values at which the device can be found after a measurement. These two values correspond to two vectors called basis vectors named as $|0\rangle$ and $|1\rangle$ and expressed as

$$|1\rangle = \begin{pmatrix} 1 \\ 0 \end{pmatrix} \qquad |0\rangle = \begin{pmatrix} 0 \\ 1 \end{pmatrix}$$

The basis vectors are unit vectors so that

$$|\langle 0|0\rangle|^2 = 1 \qquad |\langle 1|1\rangle|^2 = 1$$

Moreover, the basis vectors are mutually orthogonal, that is,

$$\langle 0|1\rangle = \langle 1|0\rangle = 0$$

One bit can store one of two possible values, e.g., either 0 or 1 corresponding to two basis vectors, i.e., $|0\rangle$ and $|1\rangle$.

The basis vectors are defined over the complex field; therefore, $|0\rangle$ and $|1\rangle$ are vectors over the multidimensional complex vector space; for example, the following two basis vectors can also represent the binary state of a bit since they are unit and orthogonal vectors:

$$\begin{pmatrix} \sqrt{2} \\ i \end{pmatrix} \qquad \begin{pmatrix} i \\ -\sqrt{2} \end{pmatrix}$$

When two bits are combined, this pair of bits can store one of four possible values, e.g., either 00, 01, 10, or 11 corresponding to four basis vectors, i.e., $|00\rangle$, $|01\rangle$, $|10\rangle$, and $|11\rangle$, and can be expressed as

$$|11\rangle = \begin{pmatrix} 1 \\ 0 \\ 0 \\ 0 \end{pmatrix} \qquad |10\rangle = \begin{pmatrix} 0 \\ 1 \\ 0 \\ 0 \end{pmatrix} \qquad |01\rangle = \begin{pmatrix} 0 \\ 0 \\ 1 \\ 0 \end{pmatrix} \qquad |00\rangle = \begin{pmatrix} 0 \\ 0 \\ 0 \\ 1 \end{pmatrix}$$

The combination of two bits expressed as four basis vectors can algebraically be expressed as the tensor product of two basis vectors as follows:

$$|11\rangle = \begin{pmatrix} 1 \\ 0 \end{pmatrix} \otimes \begin{pmatrix} 1 \\ 0 \end{pmatrix} \qquad |10\rangle = \begin{pmatrix} 1 \\ 0 \end{pmatrix} \otimes \begin{pmatrix} 0 \\ 1 \end{pmatrix} \qquad (2.1)$$

$$|01\rangle = \begin{pmatrix} 0 \\ 1 \end{pmatrix} \otimes \begin{pmatrix} 1 \\ 0 \end{pmatrix} \qquad |00\rangle = \begin{pmatrix} 0 \\ 1 \end{pmatrix} \otimes \begin{pmatrix} 0 \\ 1 \end{pmatrix} \qquad (2.2)$$

where

$$|xy\rangle = \begin{pmatrix} x_1 \\ x_2 \end{pmatrix} \otimes \begin{pmatrix} y_1 \\ y_2 \end{pmatrix} = \begin{pmatrix} x_1 \begin{pmatrix} y_1 \\ y_2 \end{pmatrix} \\ x_2 \begin{pmatrix} y_1 \\ y_2 \end{pmatrix} \end{pmatrix} = \begin{pmatrix} x_1 y_1 \\ x_1 y_2 \\ x_2 y_1 \\ x_2 y_2 \end{pmatrix} \qquad x_i \in \{0,1\} \quad y_j \in \{0,1\}$$

2.3.2 Qubits

Similar to a bit, a qubit is a device which is able to keep two states; indeed, the affix "bit" of "qubit" refers to the ability of keeping two states, whereas the prefix "qu" refers to implementing this device according to the laws of QM as depicted in Fig. 2.6. Unlike a bit (Fig. 2.6a), a qubit can store a myriad of intermediate values for a while as depicted (Fig. 2.6b). In contrast, a bit is designed not to allow this, since it is on the contrary designed to store just two values, and no intermediate values can be stored.

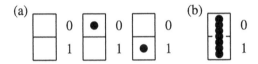

Fig. 2.6 Difference between bit and qubit. (**a**) Bit. (**b**) Qubit

The most popular implementation of a qubit is based on atoms. An atom can be in two mutually exclusive states: either the excited state or the ground state. The difference between the excited state and the ground state is due to energy. If an atom is pumped with energy and the atom is in the ground state, it will be raised to the excited state. Energy is pumped into the atom as photons (i.e., light) through a laser. When photons hit the atom, some electrons go farther from the nucleus, thus exciting the atom, that is, the atom absorbs energy. Later, this energy can be released, thus lowering the atom back down to the ground state. When an atom is neither in the excited state nor the ground state, it is in a superposed state.

Superposition can be induced by tuning the quantity of energy released from or pumped into an atom or the polarization of a photon. When an atom is hit by a moderate quantity of energy, it can be pushed into a superposed state for a non-negligible time span; a photon may also superposed between its ground state and its excited state. The state can be determined by waiting to see if it releases energy or not. If it releases energy, it goes to the ground state and a photon is emitted. If it does not release any energy, it goes to the excited state (and keeps the photon). Technically speaking, an atom with altered electrons is called an "ion." Ions can be managed by lasers when placed in isolated areas called an "ion trap." The phenomenon that energy is pumped and released through lasers hitting atoms is called "photoelectric effect."

Although qubits can physically be implemented in different ways (e.g., atoms or photons), the algebraic definition is common to every implementation of qubits, and this commonality helps reasoning about qubits without too much worrying about physical devices and properties.

Algebraically, qubits are described by vectors like bits are, that is, the state of a qubit is a vector of a multidimensional space. Recall that the affix "bit" of qubit refers to two values at which the qubit can be found after a measurement. These two values correspond to two vectors called basis vectors. Since these two values correspond to the classical bits, these two basis vectors are named as $|0\rangle$ and $|1\rangle$.

Superposition can be implemented whenever a state of a qubit that is not necessarily described by a basis vector can be represented as a linear combination of the basis vectors. A state of a qubit is therefore a vector $|\phi\rangle$ such that

$$|\phi\rangle = \alpha_0|0\rangle + \alpha_1|1\rangle$$

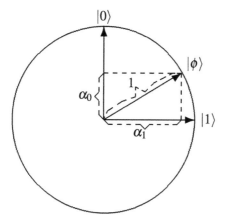

Fig. 2.7 Qubit as a vector in the real bidimensional space

where

$$0 \leq |\alpha_0|^2 \leq 1 \qquad 0 \leq |\alpha_1|^2 \leq 1 \qquad |\alpha_0|^2 + |\alpha_1|^2 = 1$$

Vectors and scalars are defined over the complex field; therefore, the αs are complex scalars, and $|0\rangle$ and $|1\rangle$ are vectors over the multidimensional complex vector space (Sect. A.4 explains why the complex field is necessary). It follows that $|\phi\rangle$ is defined over the complex field. The state $|\phi\rangle$ is said to be a superposition of the states $|0\rangle$ and $|1\rangle$, and the scalars α_0 and α_1 are said to be the amplitudes of the superposition. Figure 2.7 depicts an example of vectorial representation of a qubit in the real space.

When either $|\alpha_0|^2$ or $|\alpha_1|^2$ is 1, the qubit corresponds to a classical bit; when $|\alpha_0|^2 = 1$, then $|\alpha_1|^2 = 0$ and $|\phi\rangle = |0\rangle$; when $|\alpha_0|^2 = 0$, then $|\alpha_1|^2 = 1$ and $|\phi\rangle = |1\rangle$. It can be shown that the general qubit state vector is normalized.

The normalization of a state vector and of the amplitudes derives from the assumption that the superposition of a state with itself must result in the same state, and therefore, the sum of a state vector with itself must correspond to the same state, that is,

$$\alpha_0|\phi\rangle + \alpha_1|\phi\rangle = (\alpha_0 + \alpha_1)|\phi\rangle$$

corresponds to the same state described by $|\phi\rangle$. Thus, a state is described by the direction of the state vector, and any length of the state vector or global phase is irrelevant.

The state of two qubits combined together is the superposition of the four states corresponding to the combination of two bits. (See Sect. 2.3.1 where the four states corresponding to the combination of two bits are defined by (2.1).) The state of two qubits combined together can be generalized to the state of n qubits combined together. In this case, the state is the superposition of the 2^n states corresponding to the combination of n bits. Suppose $n = 2$. The superposition of these basis vectors is as follows:

$$|\Phi\rangle = \alpha_{00}|00\rangle + \alpha_{01}|01\rangle + \alpha_{10}|10\rangle + \alpha_{11}|11\rangle$$

where the αs are amplitudes and

$$0 \le |\alpha_{ij}|^2 \le 1 \qquad |\alpha_{00}|^2 + |\alpha_{01}|^2 + |\alpha_{10}|^2 + |\alpha_{11}|^2 = 1$$

2.4 Probability

Many IR researchers have faced the problem of defining event sets, estimating probabilities, and making predictions. For this reason, probability cannot be avoided in a book about IR. At the same time, probability is a founding topic of QM and cannot be avoided in this case too. It is therefore a little surprising that a book on the intersection between these two subjects devotes some pages to how probability is viewed from within the quantum mechanical framework applied to IR.

In this book, in particular, a special view of probability is introduced. This view derives from the quantum mechanical framework used in other scientific domains such as physics. At the level of the Dirac notation, the quantum mechanical framework may still obey the Kolmogorov axioms obeyed by classical probability, but the notation helps introduce a nonclassical probability theory when some features of QM are introduced.

2.4.1 Probability Space

A probability space is given by some observables and by a probability function of these observables. The probability function assigns a real number between 0 and 1 to each combination of observable values; for example, when an observable can be defined for the outcomes of drawing a dice, a combination of observable values contains one value only; when the dice is unbiased, the probability distribution is uniform as depicted in Fig. 2.8.

When the values of every observable are finite, the combinations of observable values are finite; for example, in IR, when two observables are defined to measure term occurrence and relevance, there are four outcomes represented by a pair of binary digits, the first digit being the outcome of the term occurrence observable and the second digit being the outcome of the relevance observable as depicted in Fig. 2.9.

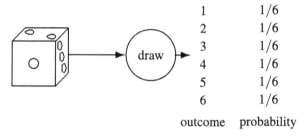

outcome	probability
1	1/6
2	1/6
3	1/6
4	1/6
5	1/6
6	1/6

Fig. 2.8 Drawing a dice

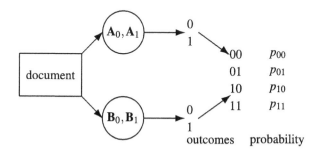

outcomes	probability
00	p_{00}
01	p_{01}
10	p_{10}
11	p_{11}

Fig. 2.9 Probability distribution of two observables

The case of observables with two values (e.g., 0 and 1) is very common in IR; term occurrence and binary relevance are two famous examples: a term either occurs ($\mathbf{B}_1$) or does not ($\mathbf{B}_0$) in a document, whereas a document is either relevant ($\mathbf{A}_1$) or not relevant ($\mathbf{A}_0$) to an information need. A term occurs in a document with a certain probability, whereas a document contains information relevant to an information need with a certain probability. A combination of two observable values i and j can be assigned a certain probability p_{ij}; for example, the probability that a relevant document contains a term can be assigned to the probability p_{11}.

More elaborate observables can be defined in IR; examples are term frequency computed at textual passage level, melodic surfaces in music documents, or pixel colors detected in images. Regardless of the degree of complexity of the observables, the conceptual structure of a probability space can still be valid.

A probability space can be represented as vectors, matrices, and operators between them. Consider an observable taking k mutually exclusive values labeled by the natural numbers $0, 1, \ldots, k - 1$; for example, this observable may be the frequency of a term within a document, the number of documents indexed in a collection, or the binary outcome of term occurrence when $k = 2$.

To each observable value, it is possible to correspond a basis vector of the k-dimensional space. Moreover, when a probability function is provided, each observable value and then each basis vector correspond to a probability measure given by the function, thus obtaining a probability distribution. We have then the following organization:

$$
\begin{array}{cccccc}
\text{Observable values} & 0 & 1 & \cdots & i & \cdots k-1 \\
\text{Basis vectors} & \begin{pmatrix} 1 \\ 0 \\ \vdots \\ 0 \\ \vdots \\ 0 \end{pmatrix} & \begin{pmatrix} 0 \\ 1 \\ \vdots \\ 0 \\ \vdots \\ 0 \end{pmatrix} & \cdots & \begin{pmatrix} 0 \\ 0 \\ \vdots \\ 1 \\ \vdots \\ 0 \end{pmatrix} & \cdots \begin{pmatrix} 0 \\ 0 \\ \vdots \\ 0 \\ \vdots \\ 1 \end{pmatrix} \\
\text{Probability values} & p_0 & p_1 & \cdots & p_i & \cdots p_{k-1}
\end{array}
$$

2.4.2 Density Matrix

The probability distribution can be arranged along the diagonal of a k-dimensional matrix called density matrix; the off-diagonal matrix elements are zeros.

$$
\begin{pmatrix}
p_0 & 0 & \cdots & 0 \\
0 & p_1 & \cdots & 0 \\
 & & \ddots & \\
0 & 0 & \cdots & p_{k-1}
\end{pmatrix}
$$

A density matrix is a Hermitian matrix and has trace 1. It is Hermitian because the conjugate transpose of the matrix is the matrix itself. It has unit trace because the diagonal elements are probability values of the same distribution and the sum of the diagonal elements is one.

The density matrix corresponding to a classical probability distribution is always diagonal and has unit trace because the sum of the diagonal elements is 1; for example, the matrix corresponding to the probability distribution of two equally probable events is

$$
\begin{pmatrix}
\frac{1}{2} & 0 \\
0 & \frac{1}{2}
\end{pmatrix}
$$

Note, however, we are exemplifying the notion of density matrix using finite and discrete probability distributions; when infinite or continuous distributions are considered, the explanation becomes more general and complex, but the meaning is essentially unaltered. Moreover, note that in QM, "density operator" is often used instead of "density matrix"; however, in this book, "matrix" is used instead of "operator."

It should be noted that not all the density matrices corresponding to a distribution need to be diagonal matrices and that the diagonal elements do not necessarily correspond to probability values, although they do have to sum up to 1; for example, the density matrix can be

$$\frac{1}{2}\begin{pmatrix} 1 & 1 \\ 1 & 1 \end{pmatrix}$$

This means that the space of the density matrices is much larger than the space of the classical probability distributions. To explain why there are more density matrices than probability distributions, first the difference between mixed distributions (or mixtures) and pure distributions needs to be introduced and then the trace rule.

A pure probability distribution is a special case of distribution. A distribution is pure when the probability is concentrated on a single event which is the certain event and has probability 1. When a distribution is not pure, it is mixed; for example, the matrix corresponding to a pure distribution of two mutually exclusive events is

$$\begin{pmatrix} 1 & 0 \\ 0 & 0 \end{pmatrix}$$

These distributions are named as "pure" to indicate the states of those objects not (yet) affected by some noise; for example, the initial state of a photon at a given polarization is pure until noise changes the polarization and the probability that the photon is at the initial state is no longer 1.

The use of "state" to mean both the particular condition that a qubit is in at a specific time and at the same time a probability distribution as described in Sect. 2.3 is not accidental. Indeed, the state of a qubit determines the probability that a given observable yields a certain value. When an atom is in a superposed state, this state is unknown but can be measured during the time span in which the atom is at a superposed state, and the probability distribution of the pure states can be estimated.

The density matrix corresponding to a pure distribution is always a projector, that is, a matrix such that the product by itself is the same matrix; for example, the pure distribution above is a projector since

$$\begin{pmatrix} 1 & 0 \\ 0 & 0 \end{pmatrix}\begin{pmatrix} 1 & 0 \\ 0 & 0 \end{pmatrix} = \begin{pmatrix} 1 & 0 \\ 0 & 0 \end{pmatrix}$$

This means that the notion of the projector is equivalent to the one of pure distribution because the former corresponds to an event that is true if and only if its probability measure is given by the pure distribution. Since a projector can correspond to one vector, the latter is called state vector.

The equivalence between pure distributions and projectors means that there are three ways to express a proposition or an event: it is possible to utilize the pure distribution corresponding to the event such that it is certain that the event occurs, to utilize the projector or to utilize the state vector; for example, suppose a photon is in its initial state x represented by a vector

$$\begin{pmatrix} 1 \\ 0 \end{pmatrix}$$

and equivalently by the projector

$$\begin{pmatrix} 1 & 0 \\ 0 & 0 \end{pmatrix}$$

Note that the vector can be viewed as a state vector (i.e., a pure density matrix) implementing a probability distribution or as a basis vector (i.e., a projector of an observable). This projector is also a pure density matrix such that whenever it is asked whether the photon is in state x, the probability will be 1 if and only if the state is precisely x. However, when x is represented by

$$\frac{1}{\sqrt{2}} \begin{pmatrix} 1 \\ 1 \end{pmatrix}$$

and equivalently by the projector

$$\frac{1}{2} \begin{pmatrix} 1 & 1 \\ 1 & 1 \end{pmatrix}$$

this density matrix will be the same as the projector, and it represents the state in which the photon is certainly in state x; it is called pure state. When a density is not pure, it is called mixed and represents a mixed state.

Pure and mixed distributions are related through a mathematical result called spectral theorem and stated below by Halmos (1987).

Theorem 2.1 (Spectral Theorem) *To every Hermitian matrix ρ on a finite-dimensional complex, inner product space corresponds real numbers $p_0, \ldots, p_{k-1}$ and projectors $\mathbf{A}_0, \ldots, \mathbf{A}_{k-1}$ so that:*

- *The p_js are pairwise distinct.*
- *The $\mathbf{A}_j$s are mutually orthogonal.*
- $\sum_{j=0}^{k-1} \mathbf{A}_j = 1.$
- $\sum_{j=0}^{k-1} p_j \mathbf{A}_j = \rho.$

The spectral theorem says that any density matrix corresponding to a distribution can be decomposed as a linear combination of projectors (i.e., pure distributions) where the eigenvalues are the probability values associated with the events represented by the projectors. Therefore, the eigenvalues are real and nonnegative and sum to 1; for example, when the matrix corresponding to the distribution of two equally probable events is considered, the spectral theorem says that

$$\rho = \begin{pmatrix} \frac{1}{2} & 0 \\ 0 & \frac{1}{2} \end{pmatrix} = \frac{1}{2}\begin{pmatrix} 1 & 0 \\ 0 & 0 \end{pmatrix} + \frac{1}{2}\begin{pmatrix} 0 & 0 \\ 0 & 1 \end{pmatrix}$$

For this reason, one speaks of mixture or mixed distribution in contrast with pure distribution—a distribution is pure when the spectral decomposition is the distribution itself or equivalently when there is a single unit eigenvalue. In contrast, when the distribution is mixed, the corresponding matrix has two or more eigenvalues which sum to 1.

The spectral theorem plays an important role in this context since it provides a probability space starting from a density matrix. That is, the spectral theorem provides an assignment to the events, which are the eigenvectors or equivalently the rank-one projectors of a density matrix, of a set of eigenvalues, which are the corresponding probabilities.

Although in classical probability every pure distribution represented by a diagonal density matrix corresponds to a projector, in general, a density matrix is not necessarily diagonal, yet it is necessarily Hermitian (i.e., symmetric in the real field); for example, consider the following density matrix:

$$\frac{1}{2}\begin{pmatrix} 1 & 1 \\ 1 & 1 \end{pmatrix}$$

It can be noted that this is a projector and has trace 1. Indeed

$$\frac{1}{2}\begin{pmatrix} 1 & 1 \\ 1 & 1 \end{pmatrix} = \frac{1}{2}\begin{pmatrix} 1 & 1 \\ 1 & 1 \end{pmatrix}\frac{1}{2}\begin{pmatrix} 1 & 1 \\ 1 & 1 \end{pmatrix}$$

Since it is a projector, it corresponds to a pure distribution. However, it is not diagonal. Since this projector corresponds to a pure distribution, there is a certain event (with probability 1) and an impossible event (with probability 0, of course) corresponding to the distribution. What is the representation of these two events? Spectral theorem provides the answer. The two events are, respectively, represented by the projectors

$$\frac{1}{2}\begin{pmatrix} 1 & 1 \\ 1 & 1 \end{pmatrix} \qquad \frac{1}{2}\begin{pmatrix} 1 & -1 \\ -1 & 1 \end{pmatrix}$$

with eigenvalues 1 and 0, respectively, or equivalently by the basis vectors

$$\frac{1}{\sqrt{2}}\begin{pmatrix} 1 \\ 1 \end{pmatrix} \qquad \frac{1}{\sqrt{2}}\begin{pmatrix} 1 \\ -1 \end{pmatrix}$$

This means that the first projector (i.e., basis vector) represents the certain event and the latter the impossible event in that probability space. It follows that the density matrix is

$$\frac{1}{2}\begin{pmatrix} 1 & 1 \\ 1 & 1 \end{pmatrix} = 1 \times \frac{1}{2}\begin{pmatrix} 1 & 1 \\ 1 & 1 \end{pmatrix} + 0 \times \frac{1}{2}\begin{pmatrix} 1 & -1 \\ -1 & 1 \end{pmatrix}$$

Although the spectral theorem provides a decomposition of a mixed distribution in a linear combination of pure distributions weighted by probabilities, this decomposition is not necessarily unique; for example,

$$\begin{pmatrix} \frac{3}{4} & 0 \\ 0 & \frac{1}{4} \end{pmatrix} = \frac{3}{4}\begin{pmatrix} 1 & 0 \\ 0 & 0 \end{pmatrix} + \frac{1}{4}\begin{pmatrix} 0 & 0 \\ 0 & 1 \end{pmatrix} = \frac{1}{4}\begin{pmatrix} \frac{3}{2} & \sqrt{3} \\ \sqrt{3} & \frac{1}{2} \end{pmatrix} + \frac{1}{4}\begin{pmatrix} \frac{3}{2} & -\sqrt{3} \\ -\sqrt{3} & \frac{1}{2} \end{pmatrix}$$

The presence of two types of distribution (i.e., mixed and pure) is due to the presence of two types of uncertainty. One uncertainty is due to the composition of an urn of elements (e.g., balls) in terms of some observables (e.g., the composition of an urn containing red balls and non-red balls); it follows that it is uncertain that an element drawn from the urn has a certain property (e.g., a ball is red) since *the urn contains elements with different values of the observable.*

Another uncertainty is due to the superposition, which was illustrated in Sect. 2.3, of the state describing each single element of the urn (e.g., the superposition of 0 and 1 in qubits); it follows that it is uncertain that an element drawn from the urn has a certain property (e.g., a ball is red) since *the element "contains" different values of the observable.*

While urn composition is the unique cause of uncertainty of classical probability, superposition is the other cause of uncertainty in quantum probability. Therefore, an urn may only contain balls (even only one) in a certain superposed state and then be described by a pure distribution, but the observation of color may still be uncertain.

The mathematical description of superposition may further explain the difference between pure distributions and mixed distributions. Consider an imaginary urn of balls. Each ball may be either red or not red. If the urn is filled with colored balls, a red ball can be drawn with a certain probability a^2. This experiment can be modeled by a mixed distribution:

$$a^2 \begin{pmatrix} 1 & 0 \\ 0 & 0 \end{pmatrix} + (1 - a^2) \begin{pmatrix} 0 & 0 \\ 0 & 1 \end{pmatrix} \tag{2.3}$$

where the projectors represent the events "red ball" and "not-red ball," respectively. Such an urn is also called "ensemble" in QM.

Alternatively, the urn may be filled with uncolored balls which become colored only when drawn from the urn. Each ball is then in a superposed state

$$|\phi\rangle = a \begin{pmatrix} 1 \\ 0 \end{pmatrix} + \sqrt{1 - a^2} \begin{pmatrix} 0 \\ 1 \end{pmatrix} \tag{2.4}$$

and becomes red with probability a^2. Therefore, there are two ways to be red: (i) a ball is (already) red when it is put into the urn and is still red when it is drawn or (ii) the ball is uncolored when it remains in the urn, and it becomes red when color is measured (i.e., the ball is drawn).

An ensemble (i.e., an urn) of uncolored balls can thus be prepared, but "redness" is a property "generated" when it is measured. Since the balls are uncolored, they may acquire a color other than red or not red if this color (e.g., black) is measured, but "blackness" is represented by another pair of projectors in the way described by the example of nonunique decomposition. Suppose a ball can be in state (2.4) with probability $\frac{1}{2}$ and in state

$$|\phi'\rangle = a \begin{pmatrix} 1 \\ 0 \end{pmatrix} - \sqrt{1 - a^2} \begin{pmatrix} 0 \\ 1 \end{pmatrix}$$

with probability $\frac{1}{2}$; note that these two state vectors represent two mutually exclusive events. The mixed distribution resulting from them is

$$\frac{1}{2} \begin{pmatrix} a^2 & a\sqrt{1-a^2} \\ a\sqrt{1-a^2} & 1-a^2 \end{pmatrix} + \frac{1}{2} \begin{pmatrix} a^2 & -a\sqrt{1-a^2} \\ -a\sqrt{1-a^2} & 1-a^2 \end{pmatrix} = \begin{pmatrix} a^2 & 0 \\ 0 & 1-a^2 \end{pmatrix}$$

which is the mixed distribution when the balls are already red or not before measurement.

The basic difference between the notion of pure distribution and the notion of mixed distribution is that a pure distribution models the uncertainty of an individual element of a set, whereas a mixed distribution models the uncertainty of the whole set, that is, it models the frequency of the values of an observable when it is measured in the elements of the set. In contrast, when an individual element is measured, the uncertainty of the measurement of an observable in the element can be modeled by a pure distribution, which is referred to the element and not to the set.

Mixed and pure distributions can be combined in one single state. Suppose an urn of uncolored balls is prepared. The balls can be prepared in two different superposed states, say, $|\phi_1\rangle$ and $|\phi_2\rangle$. A mixture of balls is then present in the urn and can be represented by the following mixed density matrix:

$$q_1 |\phi_1\rangle\langle\phi_1| + q_2 |\phi_2\rangle\langle\phi_2|$$

where q_i is the probability that a ball is in state $|\phi_i\rangle$. Consider the observable of a color (e.g., red) of a ball represented by the projectors (2.3). The probability that a ball drawn from the urn is red is then

$$q_1 |\langle 1|\phi_1\rangle|^2 + q_2 |\langle 1|\phi_2\rangle|^2$$

If the urn is filled with some uncolored balls in state $|\phi\rangle$ and some red balls in state $|1\rangle$, the uncertainty of drawing a red ball can be represented by the following mixed density matrix:

$$q_1|\phi\rangle\langle\phi| + q_2|1\rangle\langle1|$$

The extension of the algebraic form to the case of a bidimensional probabilistic space ($d = 2$) is quite straightforward. In general, any pair of event sets can be considered; an event is an observable value or a given combination of observable values. Suppose that both event sets are written using $\{0, 1\}$. When an event is observed from each set, there are four possibilities: 00, 01, 10, and 11 where the first bit refers to the first event set and the second bit refers to the second event set.

In general, the product of d binary sets is an event set of size 2^d. Hence, the probability distribution has four values $p_{00}, p_{01}, p_{10}, and\ p_{11}$ which can be arranged along the diagonal of a density matrix (0 elsewhere) such that p_{ij} is the probability that the outcome from the first event set is i and that from the second event set is j. Consider, for example, two terms such that either they co-occur or do not with equal probability. The matrix corresponding to this distribution is

$$\frac{1}{2}\begin{pmatrix} 1 & 0 & 0 & 0 \\ 0 & 0 & 0 & 0 \\ 0 & 0 & 0 & 0 \\ 0 & 0 & 0 & 1 \end{pmatrix} \tag{2.5}$$

where the top-left element, $p_{00} = \frac{1}{2}$, is the probability that neither term occurs and the bottom-right element, $p_{11} = \frac{1}{2}$, is the probability that both terms occur. In general, any pair of events can have the probability distribution arranged along the diagonal of (2.5); other examples of pairs of events are term occurrence and relevance, aboutness and relevance, and document retention and term occurrence.

The marginal distributions can be computed using the usual rules, that is,

$$f_0 = p_{00} + p_{01} \qquad f_1 = p_{10} + p_{11}$$

$$g_0 = p_{00} + p_{10} \qquad g_1 = p_{01} + p_{11}$$

where f is the marginal distribution of the first event set and g is the marginal distribution of the second event set. Both marginal distributions can be arranged along a diagonal density matrix as follows:

$$\begin{pmatrix} f_0 & 0 \\ 0 & f_1 \end{pmatrix} \qquad \begin{pmatrix} g_0 & 0 \\ 0 & g_1 \end{pmatrix}$$

Suppose that the distribution of the product of two event sets is uncorrelated (i.e., the events are independent). It follows that

$$p_{ij} = f_i g_j \qquad i, j = 0, 1$$

A distribution p is uncorrelated when it can be written as a tensor product of f and g. If f and g can be written as diagonal matrices, the tensor product is

$$\begin{pmatrix} p_{00} & 0 & 0 & 0 \\ 0 & p_{01} & 0 & 0 \\ 0 & 0 & p_{10} & 0 \\ 0 & 0 & 0 & p_{11} \end{pmatrix} = \begin{pmatrix} f_0 & 0 \\ 0 & f_1 \end{pmatrix} \otimes \begin{pmatrix} g_0 & 0 \\ 0 & g_1 \end{pmatrix}$$

A correlated distribution cannot be written as a tensor product, that is, the event sets are not independent; for example, $(\frac{1}{2}, 0, 0, \frac{1}{2})$ is correlated; indeed, both marginals are $(\frac{1}{2}, \frac{1}{2})$, and one can see that their tensor product is

$$\frac{1}{2}\begin{pmatrix} 1 & 0 \\ 0 & 1 \end{pmatrix} \otimes \frac{1}{2}\begin{pmatrix} 1 & 0 \\ 0 & 1 \end{pmatrix} = \frac{1}{4}\begin{pmatrix} 1 & 0 & 0 & 0 \\ 0 & 1 & 0 & 0 \\ 0 & 0 & 1 & 0 \\ 0 & 0 & 0 & 1 \end{pmatrix} \neq \frac{1}{2}\begin{pmatrix} 1 & 0 & 0 & 0 \\ 0 & 0 & 0 & 0 \\ 0 & 0 & 0 & 0 \\ 0 & 0 & 0 & 1 \end{pmatrix}$$

A pure distribution is uncorrelated. Indeed, there exists only one pair i, j such that $p_{ij} = 1$. It follows that $f_i = 1$ and $g_j = 1$ which are the only values such that $p_{ij} = f_i g_j$ for every i, j. This can be seen using tensor products, for example:

$$\begin{pmatrix} 1 & 0 & 0 & 0 \\ 0 & 0 & 0 & 0 \\ 0 & 0 & 0 & 0 \\ 0 & 0 & 0 & 0 \end{pmatrix} = \begin{pmatrix} 1 & 0 \\ 0 & 0 \end{pmatrix} \otimes \begin{pmatrix} 1 & 0 \\ 0 & 0 \end{pmatrix}$$

2.4.3 Trace Rule

It may have been noted that the basis vectors of an observable have length 1. This constraint is required due to the way the probability of an event is computed. When using this algebraic form to represent probability spaces, the function for computing a probability is the trace of the matrix obtained by multiplying the density matrix with the projector corresponding to the event. The usual notation for the probability of the event represented by projector $\mathbf{A}$ when the distribution is represented by density matrix ρ is

$$\mathrm{tr}(\rho \mathbf{A})$$

For example, when

$$\rho = \frac{1}{2}\begin{pmatrix} 1 & 0 \\ 0 & 1 \end{pmatrix} \quad \text{and} \quad \mathbf{A} = \begin{pmatrix} 1 & 0 \\ 0 & 0 \end{pmatrix}$$

the probability of the event represented by $\mathbf{A}$ is

$$\text{tr}(\rho\mathbf{A}) = \text{tr}\left[\begin{pmatrix} \frac{1}{2} & 0 \\ 0 & \frac{1}{2} \end{pmatrix}\begin{pmatrix} 1 & 0 \\ 0 & 0 \end{pmatrix}\right] = \text{tr}\begin{pmatrix} \frac{1}{2} & 0 \\ 0 & 0 \end{pmatrix} = \frac{1}{2}$$

When $\mathbf{A} = |x\rangle\langle x|$ is a projector, the trace-based probability function can be written as

$$\text{tr}(\rho\mathbf{A}) = \langle x|\rho|x\rangle$$

When ρ is a projector $|y\rangle\langle y|$, then

$$\text{tr}(\rho\mathbf{A}) = |\langle x|y\rangle|^2 \tag{2.6}$$

where $\langle x|y\rangle$ is called probability amplitude or amplitude in short; an explanation of why amplitudes are squared instead of, say, cubed or raised to an arbitrary power will be given in Sect. A.2; Fig. 2.10 depicts the (simple) case of (2.6) in the real space where the amplitude is just a coordinate in a vector space.

In general, an amplitude is a complex number, whereas the squared modulus of the amplitude is a real number, the latter being a necessary condition that the trace rule yields probabilities.

Using the spectral theorem, it is possible to express the probability of the event (or observable value) represented by a projector $\mathbf{A}$ as a function of a given density matrix. Suppose ρ is a density matrix. According to the spectral theorem, we have that

$$\rho = p_0\mathbf{A}_0 + \cdots + p_{k-1}\mathbf{A}_{k-1}$$

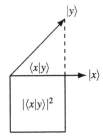

Fig. 2.10 Trace rule

Suppose $\mathbf{A} = |x\rangle\langle x|$ is a projector that represents an observable value. The probability of this event is provided by the trace rule

$$\mathrm{tr}(\rho\mathbf{A})$$

As the trace is a linear function, we have that

$$
\begin{aligned}
\mathrm{tr}(\rho\mathbf{A}) &= \mathrm{tr}((p_0\mathbf{A}_0 + \cdots p_{k-1}\mathbf{A}_{k-1})\mathbf{A}) \\
&= \mathrm{tr}(p_0\mathbf{A}_0\mathbf{A} + \cdots p_{k-1}\mathbf{A}_{k-1}\mathbf{A}) \\
&= \mathrm{tr}(p_0\mathbf{A}_0\mathbf{A}) + \cdots \mathrm{tr}(p_{k-1}\mathbf{A}_{k-1}\mathbf{A}) \\
&= p_0\mathrm{tr}(\mathbf{A}_0\mathbf{A}) + \cdots p_{k-1}\mathrm{tr}(\mathbf{A}_{k-1}\mathbf{A}) \\
&= p_0\mathrm{tr}(|y_0\rangle\langle y_0|x\rangle\langle x|) + \cdots p_{k-1}\mathrm{tr}(|y_{k-1}\rangle\langle y_{k-1}|x\rangle|x\rangle) \\
&= p_0\langle x|y_0\rangle\langle x|y_0\rangle + \cdots p_{k-1}\langle x|y_{k-1}\rangle\langle x|y_{k-1}\rangle) \\
&= p_0|\langle y_0|x\rangle|^2 + \cdots + p_{k-1}|\langle y_{k-1}|x\rangle|^2
\end{aligned}
$$

The last line tells us that the probability of the event represented by the projector $\mathbf{A}$ is a linear combination of probabilities expressed as

$$p_j|\langle y_j|x\rangle|^2 \qquad j = 0, \ldots, k-1$$

where p_j is the probability of the event represented by $\mathbf{A}_j$ and $|\langle y_j|x\rangle|^2$ is the probability of the event represented by $\mathbf{A}$ conditioned to the event represented by the pure distribution (or projector) $\mathbf{A}_j$.

This is the reason why a density matrix is called state. The density matrix of a pure distribution is the state of the object on which observables are applied; when an object is a given state, its pure distribution is given by a density matrix which is also a projector, i.e., the projector of the event that the object is in that state. When the distribution is pure, the state is pure too. In general, the state of an object is not pure; it is represented by a density matrix which is a mixture of pure distributions weighted by $p_0, \ldots, p_{k-1}$.

The expression $\mathrm{tr}(\rho\mathbf{A})$ of the probability of the event represented by $\mathbf{A}$ when the object is in the state given by the density matrix ρ in terms of the weighted sum of probabilities, that is,

$$p_0|\langle y_0|x\rangle|^2 + \cdots p_{k-1}|\langle y_{k-1}|x\rangle|^2$$

is based on the square rule, that is, the rule for which the probability that x is observed when the state is y is $|\langle y|x\rangle|^2$. The square rule is a specific case of the trace rule; the former holds for general density matrices and the latter holds when the density matrix corresponds to a state vector; see also Sect. A.2.

2.5   Interference

The best way to introduce interference is to refer to the lectures of Feynman et al. (1965) who explain this concept by blending mathematical abstraction and physical implementation without exceeding in either of these two extremes.

2.5.1   Double-Slit Experiment

Feynman et al. (1965) introduced an example which is currently the most used example in the relevant literature; this example is based on a system composed of three main components (Fig. 2.11a), that is, a source of particles (e.g., electrons) and a wall with two slits placed between the source and a shield equipped with a series of detectors of particles.

The source emits particles which are supposed to travel in the space from the source toward the wall. When a particle arrives at the wall, it might pass or not pass through a slit in the wall; the slits are supposed to be small enough to allow only one single particle to pass through the slit at a time. Although not all the particles sent from the source pass through a slit in the wall, the number of particles passed through a slit will be high enough to calculate a distribution of the frequencies of

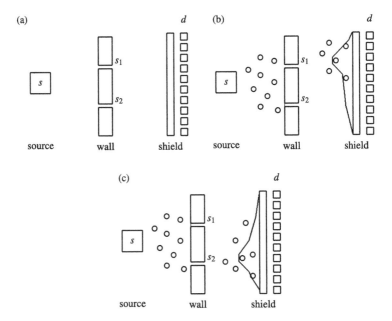

Fig. 2.11 Double-slit experiment

the particles arriving at the detector; if a large number of particles are sent to the wall, the number of particles that pass through a slit in the wall is not negligible.

It is common to suppose that the source that emits the particles is a sort of gun firing particles in all directions. When the particles are emitted in all directions, they can arrive at a slit from a variety of directions which are therefore not all perpendicular to the slit. In this situation, the particles that pass through a slit can take any possible direction which are not all perpendicular to the slit. Moreover, the particles that do not pass exactly through the center of a slit may touch an edge of the slit and change their own direction toward a point of the detector not perpendicular to the center of the slit.

The shield can be equipped with a series of detectors of particles which can count the particles arriving from a slit. In the shield, there are some detectors in front of and close to the slits, and there are some other detectors far from the slit; the detectors can be uniformly distributed in the shield, that is, they can be placed at an equal distance from each other.

Suppose one slit is closed and the other is open. Experiments have shown that the distribution of the frequencies of the particles arriving at the shield and counted by the detectors of the shield are bell shaped because the detector placed in front of the slit counts most of the particles passing through the slit, the detectors closest to the detector placed in front of the slit count many other particles, and the frequency of particles counted decreases the further the detectors that counted them are far from the detector placed in front of the slit (Fig. 2.11b).

When the slit that is currently open is closed and the closed slit is opened, the distribution of frequency of the particles arriving at the shield and counted by the detectors of the shield is again bell shaped, but the curve will be translated to the open slit (Fig. 2.11c).

It is possible to give the algebraic description of the events just described and of the related probabilities of the system consisting of the source of particles, the wall, and the shield. Suppose s is the state of the particle when leaving the source, s_1 is the state of the particle when it is passing through slit 1, and s_2 is the state of the particle when it is passing through slit 2.

In QM, states and events are described by vectors defined over the complex field. Suppose d is the event that a detector detects (and counts) a particle passing through an open slit. The inner product between a state vector and a vector of an event yields a complex number which is a probability amplitude; for example, when the state is s_1, the probability amplitude of the event that a particle is detected by d is $\langle d|s_1 \rangle$.

In QM, the probability that a particle is detected by d when the state is s_1 is the squared probability amplitude, that is,

$$P(\text{a particle is detected by } d \text{ when the state is } s_1) = |\langle d|s_1 \rangle|^2$$

This probability can be estimated using the classical rules of statistics, that is, the empirical probability that a particle is detected by d when the state is s_1 can be estimated by the relative frequency of the number of particles detected by d which passed through s_1.

Suppose the slits are continuously opened and closed in such a way that when a slit is open, the other is closed. We can easily imagine a device which opens and closes the slits at regular time intervals while guaranteeing the mutual exclusiveness of the opening. If this device operates for a long enough time, the fraction of instants a slit is open is about $\frac{1}{2}$, which is also the probability that a slit is open. When this device is in operation and the source is emitting the particles as if it were a gun shooting in all directions, the particles can pass through the open slit and cannot of course pass through the closed slit. In this situation, the probability that a particle is detected by d is simply the weighted sum of the probability that the particle is detected by d when s_1 is open and of the probability that the particle is detected by d when s_2 is open; the weights of this sum are given by the fraction of instants a slit is open. We can therefore write that

$$P(\text{a particle is detected by } d) = \frac{1}{2}|\langle d|s_1\rangle|^2 + \frac{1}{2}|\langle d|s_2\rangle|^2 \tag{2.7}$$

Note that the formulation given by (2.7) implies that the state that a particle has passed through a slit s_j can be represented by the pure density matrix $|s_j\rangle\langle s_j|$, thus stating that (2.7) can be written as

$$\frac{1}{2}\text{tr}(|s_1\rangle\langle s_1|d\rangle\langle d|) + \frac{1}{2}\text{tr}(|s_2\rangle\langle s_2|d\rangle\langle d|)$$

where $|d\rangle\langle d|$ is the projector of the event observed.

This result is nothing but the consequence of the distributive law applied to the event d when the states s_1 and s_2 are viewed as events. When observing the event d, one is detecting a particle under the condition that either slit 1 or slit 2 is open, that is, one is evaluating the event d. This event can be intersected with the event that is always true, i.e., the event that a slit is open; indeed, the device that is opening and closing the slits guarantees that a slit is always open. It follows that we can write

$$d = d \wedge (s_1 \vee s_2)$$

where $s_1 \vee s_2$ is always true although only, and because certainly, one slit is open. The distributive law applied to d allows us to "distribute" d across the events written between the parentheses and to obtain

$$d = (d \wedge s_1) \vee (d \wedge s_2) \tag{2.8}$$

As we have assumed that either s_1 or s_2 is open and that both slits cannot simultaneously open, that is, s_1 and s_2 are mutually exclusive events, the particle detected by d arrived either from s_1 or s_2. Therefore, it is possible to calculate the probability that a particle is detected by d as follows:

$$P(\text{a particle is detected by } d) = P(d \wedge s_1) + P(d \wedge s_2) \tag{2.9}$$

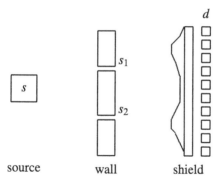

Fig. 2.12 Mixed distribution of the double-slit experiment

The probabilities on the right-hand side of the latter expression can be written using the Bayes postulate to obtain

$$P(\text{a particle is detected by } d) = P(d|s_1)P(s_1) + P(d|s_2)P(s_2)$$

As we have stated that $P(s_1) = P(s_2) = \frac{1}{2}$ provided the mode of operation of the device controlling the slits, the latter expression becomes (2.7) since $P(d|s_1) = |\langle d|s_1\rangle|^2$ and $P(d|s_2) = |\langle d|s_2\rangle|^2$. If the distribution of the frequencies of the particle were calculated, something like the plot in Fig. 2.12 would be observed.

2.5.2 Interference Term

So far we have assumed that either s_1 or s_2 is open and that both slits cannot simultaneously be open. This assumption implies that the particle that is detected by d arrived either from s_1 or s_2. The question is, what happens when both slits are simultaneously open? Intuition suggests that a particle that is detected by d still arrives from either s_1 or s_2 since it is unnatural that a particle passes through bits s_1 and s_2 at the same time. If this intuition were correct and is supported by the experiments, the distributive law should still be applied, (2.7) would still be valid, and the plot in Fig. 2.12 would be observed.

However, experiments showed that this is not the case and that (2.7) is on the contrary invalid; in contrast, the probability that a particle is detected by d when both slits are open is described by another law, and a pattern like the plot in Fig. 2.13 will be observed.

In QM, the event that a particle is detected by d when both slits are open is still described by the projector $|d\rangle\langle d|$ as it is done when either s_1 or s_2 is open, but the state vector $|s\rangle$ is a superposition of the state vectors $|s_1\rangle$ and $|s_2\rangle$, that is, the state that both slits are open translates into a superposition of states.

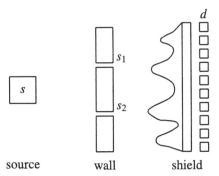

Fig. 2.13 Superposed distribution of the double-slit experiment

To understand how this superposition can be built, it is necessary to introduce the notation of the path along which a particle travels from the source through the slits and to the detectors. To this end, consider for a while the path described by (2.8) and measured by (2.9). When a particle can reach a detector by two possible slits, the total probability for this process is the sum of the probabilities for the two paths considered separately; this is a consequence of the mutual exclusiveness of the paths passing through the slits.

Consider the quantum mechanical framework instead. In this framework, we have to reason using probability amplitudes and not probabilities. When a particle can reach a detector by two possible slits, the total probability amplitude for this process is the sum of the probability amplitudes for the two paths considered separately, that is,

$$\langle d|s \rangle = \langle d|s_{\text{through slit 1}} \rangle + \langle d|s_{\text{through slit 2}} \rangle$$

Consider the first term on the right-hand side. This term is the amplitude of the path traveled by a particle leaving the source, passing through slit s_1, and arriving at the detector d. When a particle goes by some particular path, the probability amplitude of that path can be written as the product between the probability amplitude of going from the beginning of the path up to a given point and the probability amplitude of going from that point to the end of the path. If that point is slit s_1, we have that the probability amplitude of the path from the source to the detector can be written as the product of the probability amplitude of going from the source to s_1 and of the probability amplitude of going from s_1 to the detector:

$$\langle d|s_{\text{through slit 1}} \rangle = \langle d|s_1 \rangle \langle s_1|s \rangle$$

Using the same reasoning for the path passing through the other slit, we have that

$$\langle d|s_{\text{through slit 2}} \rangle = \langle d|s_2 \rangle \langle s_2|s \rangle$$

Note the similarity with the classical probability rule chain, that is,

$$P(d|\text{start from } s \text{ through slit } i) = P(d|s_i)P(s_i|s)$$

The probability amplitude of the path traveled by a particle leaving the source, passing through slit 1, and arriving at the detector d is therefore

$$\langle d|s\rangle = \langle d|s_1\rangle\langle s_1|s\rangle + \langle d|s_2\rangle\langle s_2|s\rangle$$

The passage from probability amplitude to probability is dictated by the trace rule. According to this rule, the probability of the path traveled by a particle leaving the source, passing through slit 1, and arriving at the detector d is

$$|\langle d|s\rangle|^2$$

that is,

$$|\langle d|s_1\rangle\langle s_1|s\rangle + \langle d|s_2\rangle\langle s_2|s\rangle|^2$$

The latter is the squared modulus of a complex number resulting from the sum of two complex numbers. It is known that if z is a complex number, then $|z|^2 = zz^*$. Moreover, if z_1, z_2 are two complex numbers, then

$$
\begin{aligned}
|z_1 + z_2|^2 &= (z_1 + z_2)(z_1 + z_2)^* \\
&= (z_1 + z_2)(z_1^* + z_2^*) \\
&= z_1 z_1^* + z_1 z_2^* + z_2 z_1^* + z_2 z_2^* \\
&= |z_1|^2 + 2|z_1||z_2| + |z_2|^2
\end{aligned}
$$

Let

$$z_1 = \langle d|s_1\rangle\langle s_1|s\rangle \qquad z_2 = \langle d|s_2\rangle\langle s_2|s\rangle$$

We have that

$$|z_1|^2 = |\langle d|s_1\rangle|^2|\langle s_1|s\rangle|^2$$

and

$$|z_2|^2 = |\langle d|s_2\rangle|^2|\langle s_2|s\rangle|^2$$

It follows that

$$|\langle d|s\rangle|^2 = |a_1|^2|\langle d|s_1\rangle|^2 + |a_2|^2|\langle d|s_2\rangle|^2 + I \qquad (2.10)$$

where

$$a_1 = \langle s_1|s \rangle \qquad a_2 = \langle s_2|s \rangle$$

are amplitudes and

$$I = 2|\langle d|s_1 \rangle \langle s_1|s \rangle \langle s|s_2 \rangle \langle s_2|d \rangle| \cos \theta$$

where θ is the angle of the complex number $\langle d|s_1 \rangle \langle s_1|s \rangle \langle s|s_2 \rangle \langle s_2|d \rangle$. Equation (2.10) shows that $|s \rangle$ is indeed a superposition of two state vectors $|s_1 \rangle$ and $|s_2 \rangle$, i.e., a sum of two vectors weighted by two probability amplitudes, that is,

$$|s \rangle = a_1|s_1 \rangle + a_2|s_2 \rangle$$

where $a_1 = \langle s_1|s \rangle$ and $a_2 = \langle s_2|s \rangle$.

An immediate consequence of the fact that the probability amplitude that a particle is detected by d when the state is s becomes the sum of two probability amplitudes is the inapplicability of the distributive law to the events described by $|s_1 \rangle$, $|s_2 \rangle$, and $|d \rangle$. More precisely, the vector $|d \rangle$ can still be distributed in the sum of $|s_1 \rangle + |s_2 \rangle$, but this distribution of probability amplitudes does not correspond to a distribution of probabilities. The inapplicability of the distributive law is one of the most striking facts of QM.

I is called *interference term*, and it is not only an algebraic accident; it is a necessary component of the probabilistic model describing the distribution of frequency of the particles detected by the detectors of the shield when both slits are open.

In other words, the experiments confirmed that (2.10) is indeed the probability that a particle is detected by d when both slits are open. To understand why this term describes a sort of interference, consider the expression of I. This is a product of probability amplitudes. A product of probability amplitudes is used to give the probability amplitude of the path traveled by a particle which in this case goes from s to d. Consider the center of the expression of I. From s, a particle travels to slit 1 and to slit 2. The meaning of the product $\langle s|s_1 \rangle$ of $|s \rangle$ by $\langle s_1|$ and of the product $\langle s|s_2 \rangle$ between $|s \rangle$ and $\langle s_2|$ is indeed the fact that the particle goes to both slits as if it were traveling two paths at the same time. The remainder of the expression of I, which ends in $|d \rangle$, is still based on the simultaneity of the trajectories of the particle. The cosine of the angle θ ranges from -1 to $+1$, and therefore, I ranges from -2 to $+2$. Therefore, $|\langle d|s \rangle|^2$ might be less or greater than $|a_1|^2|\langle d|s_1 \rangle|^2 + |a_2|^2|\langle d|s_2 \rangle|^2$, and I gives the sinusoidal shape to the plot in Fig. 2.13.

2.6 Entanglement

Entanglement refers to the state of a pair of particles (e.g., photons) twisted together in such a way that the value of an observable of a particle is at every instant of time equal to the value of the observable of the other particle; for example, the polarization of a particle is always equal to the polarization of the other particle when the polarization changes; a full account is reported by Zeilinger (2010).

2.6.1 Alice and Bob

Alice and Bob are two imaginary characters. They have been often utilized to introduce entanglement in the textbooks about QM. Their names were chosen because the initials are, respectively, A and B, the latter being the labels used to name variables; actually, other characters such as Charlie and Eve enter into play when other variables are needed to describe other mechanisms of QM. Alice and Bob live in two distant places; actually, the distance between them is not really crucial from a narrative point of view; however, it will be important when explaining the strangeness of entanglement.

Somebody else prepares a sequence of pairs of photons in a special way using a generator of pairs of entangled photons, and for each pair of photons, it sends one photon of the pair to Alice and the other photon of the pair to Bob. The state of each pair of photons is entangled in the sense that both photons have the same oscillation, e.g., either they have an "up" orientation or they have a "down" orientation.

Then, Alice receives one photon of the pair and Bob receives the other photon; since we need to name the photons, let ϕ_A be the photon received by Alice and ϕ_B be the photon received by Bob. Alice and Bob cannot exchange any information when they will be called to operate on the photons which have been given to each of them. Therefore, the two characters operate in an isolated environment, and the only tie between them is constituted by the entanglement of the photons.

Finally, Alice and Bob are provided with one Polarizing Beam Splitter (PBS) each; they use their own splitter to measure the photon sent to them. Before performing the measurement, they set the PBS to an oscillation so that the splitter can measure the oscillation and output one orientation; for example, the splitter is set to measure the vertical oscillation and output either "up" (i.e., $+1$) or "down" (i.e., -1).

If Alice measures the vertical oscillation of ϕ_A and obtains $+1$, then the state of the photon becomes "up," and the state of Bob's photon is also "up" since the photons are entangled. So the state of the pair of photons will be "up-up." If Bob measures his photon, he will observe "up." In a similar way, if Alice measures "down," Bob will measure the same. When measuring in the vertical oscillation, Alice and Bob will always observe the same outcomes.

Alice and Bob are located in distant places and cannot communicate. Moreover, the outcome of the measurement of one photon of a pair is random; indeed, Alice sees a random sequence of -1s and $+1$s, and Bob sees the same; however, he cannot know in advance the outcome observed by Alice. Therefore, the only explanation of the perfect agreement between the two measurements can be entanglement. Nevertheless, entanglement is a characteristic of the photons and not of the observed oscillations. The fact that the photons are entangled does not imply that they are also interacting, that is, that they are acting between them exchanging information or that the PBSs are operating by exchanging input and output.

The experiments performed with entangled photons also showed other interesting statistical outcomes. When Alice and Bob prepare their PBSs at different settings, the distribution of the frequencies of the outcome follows a certain probability function which depends on the difference between the settings. When Alice prepares her PBS to measure vertical oscillation and Bob prepares his PBS to measure an oblique oscillation such that the difference between the oscillations is $30°$, the proportion of pairs of photons such that the outcomes of the measurements coincide is about 75 %, and the proportion of pairs of photons such that the outcomes of the measurements do not coincide is about 25 %. When the difference between the oscillations is $60°$, the proportion of pairs of photons such that the outcomes of the measurements coincide is about 25 %, and the proportion of pairs of photons such that the outcomes of the measurements do not coincide is about 75 %. In general, when the difference is $\theta°$, the proportion of pairs of photons such that the outcomes of the measurements coincide is about $\cos^2 \theta \times 100$ %, and the proportion of pairs of photons such that the outcomes of the measurements do not coincide is about $1 - \cos^2 \theta \times 100$ %.

2.6.2 Local Hidden Variables and Bell's Inequality

The question is why we should be certain that neither photons nor PBSs interact. The answer has been provided by different experiments which placed the photons at large distances between them and prepared the PBSs in a way that the measurement of a property (e.g., oscillation) took place at the same instant. The difference between the instants was so small, and the distance was so large that the speed which would have been needed to transmit some information from a photon or PBS to another photon or PBS should have been greater than the speed of light, the latter being an impossible condition.

The only possibility that the photons can exchange information might be the presence of local-hidden variables of the photons other than those measured by the PBSs in the photons themselves. These variables would be "onboard" on photons and would instruct the PBSs to output the same outcome whenever they measure the same oscillation in both photons. These variables are labeled as "local and hidden" since they cannot be directly observed (i.e., they are hidden) and they are internal (i.e., local) to each of the photons and do not depend on measurements performed by faraway devices; they act as the genetic attributes inherited by twins who may manifest the same characteristics over time. Although the local-hidden variables cannot be observed, they might actually exist and determine what the PBSs outputs. Therefore, the random sequence of orientations observed by Alice and Bob might be caused by the random behavior of the local-hidden variables.

However, it is not possible to construct a theory based on local-hidden variables that agrees with the experimental results. Bell (1964) explained that any theory based on local-hidden variables must produce experimental results which satisfy a statistical inequality called Bell's inequality. When the experimental results observed in entangled photos are used to estimate the quantities involved in Bell's inequality, the latter is violated, thus concluding that a theory based on local-hidden variables is incompatible with entanglement. Then, Bell proved that the experimental results based on the measurements of photon oscillations are incompatible with a theory based on local-hidden variables when performed on entangled photons.

2.7 Detection

Detection consists of identifying the information concealed in the data transmitted by the source placed on one side to the detector placed on the other side. The key problem of detection is that the data transmitted through a channel are only an approximation of the "true" information that one side wants to transmit; for example, in IR, the information that is relevant to the user's information need is transmitted by a system to the user by means of a document which is only an approximation of the information fulfilling the user's need.

The standard configuration of a communication system consists of a source, a channel, and a detector as depicted in Fig. 2.14. The source emits digital signals a ("true" information) chosen from a fixed alphabet; for example, the alphabet can be binary, and the signal emitted is either 0 or 1. The signals are sent to a receiver through a channel, and the receiver measures the symbol x which may differ from the original signal a because of noise or distortion.

In quantum detection, there is a coder between the source and the channel as depicted in Fig. 2.15. The coder encodes the signal into a particle and assigns to the particle a pure state ϕ; the particle is thus described by its pure state vector $|\phi\rangle$. As each signal of a fixed finite alphabet is assigned a prior probability of emission

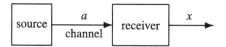

Fig. 2.14 Standard communication system

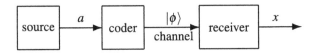

Fig. 2.15 Quantum communication system

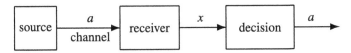

Fig. 2.16 Detection as a decision problem

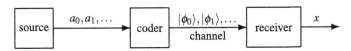

Fig. 2.17 Two symbols sent through a quantum channel

and the coder does not intervene on the source, each pure state has its own prior probability equal to the prior probability of the signal.

Once the particle is received on the other side, the receiver performs a measurement on the particle transmitted through the channel. This measurement is accomplished by an observable which usually yields as many values as the number of distinct pure states; for example, if there are two possible pure states, the detector may detect two values. However, in general, the number of values may be different than the number of pure states; for example, there might be two pure states and four observable values.

The values observed can serve to decide about the original signal and then about the pure state of the signal given by the coder. The decision taken depends on the region of values to which an observed value belongs. If the observed value x belongs to a certain region of acceptance, the overall system decides that the original signal was, say, a_0; otherwise, it was a_1 as depicted in Fig. 2.16; for example, if $x = 0$ is detected, the system decides a_0; otherwise, it decides a_1.

2.7.1 Detection, Projectors, and Probability

Algebraically, the setting above can be described using state vectors, density matrices, and projectors. It is known that the pure state of a signal can be represented as a state vector which belongs to a vector basis together with the other basis vectors corresponding to the possible signals and pure states. Suppose the alphabet is binary and consists of only two signals. These signals are assigned to the pure state vectors $|\phi_0\rangle$ and $|\phi_1\rangle$ as depicted in Fig. 2.17. Using the trace rule to compute probabilities, it is possible to compute the probability that the detector receives a symbol given a

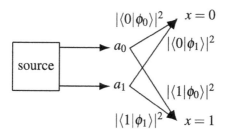

Fig. 2.18 Communication channel with probabilities

state, that is,

$$P(x \text{ is received when the state is } \phi_j) = |\langle x|\phi_j\rangle|^2$$

where $|x\rangle$ is the basis vector of the symbol x measured by the detector. Figure 2.18 depicts the network of signals, probabilities, and symbols of the channel where the q_is are the prior probability that a_i is sent by the source.

The construction of the observable used to detect the state of the particle sent through the channel is a crucial step since the correctness of the decision and the related probability of error of decision depend on the effectiveness of the observable values in supporting the decision. As an observable consists of values, it is necessary to define the values to be observed. The decision about the state of the particle consists of partitioning the set of observable values so that each part of this set corresponds to a state. Once the set of observable values is partitioned, the probability that a state is decided when a given state was originally set and the particle has been sent can be computed as follows:

$$p(\phi_i|\phi_j) = P(\text{the final decision is } \phi_i \text{ when the original state is } \phi_j)$$

For example, in the event of a set of binary observable values and of two pure states, the value $x = 0$ may correspond to ϕ_0 and the value $x = 1$ may correspond to ϕ_1; in the event of a set of four observable values, the values $x \in \{0, 1\}$ may correspond to ϕ_0 and the value $x \in \{2, 3\}$ may correspond to ϕ_1 as depicted in Fig. 2.19. The channel is associated with a probability network measuring the degree to which the signals are observed; when measuring the channel, a symbol $x \in \{0, 1, 2, 3\}$ is observed and used to decide whether the symbol emitted was a_0 or a_1; to this end, the set of observable values is split in two distinct regions; one region $\{0, 1\}$ is associated with a_0, and the other region $\{2, 3\}$ is associated with a_1 as depicted in Fig. 2.19.

The final decision ϕ_i passes through the observed values x which are in turn generated in a given state ϕ_j with the probability $|\langle x|\phi_j\rangle|^2$. Suppose A_i is the region of observable values leading to decide for ϕ_i. When the set of observable values is

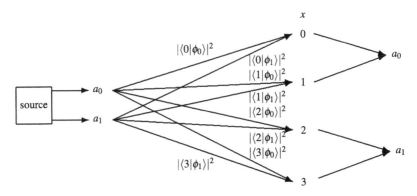

Fig. 2.19 Communication channel with probabilities and decision

finite and discrete, we have that

$$p(\phi_i|\phi_j) = \mathrm{P}(x \in A_i \text{ when the original state is } \phi_j) = \sum_{x \in A_i} |\langle x|\phi_j \rangle|^2$$

As each x corresponds to a basis vector $|x\rangle$ and then to a projector $|x\rangle\langle x|$ mutually orthogonal to the other projectors of the observable values, it is possible to write as follows:

$$\sum_{x \in A_i} |\langle x|\phi_j \rangle|^2 = \sum_{x \in A_i} \mathrm{tr}(|x\rangle\langle x|\phi_j\rangle\langle\phi_j|)$$

$$= \mathrm{tr}\left(\left(\sum_{x \in A_i} |x\rangle\langle x|\right) |\phi_j\rangle\langle\phi_j|\right)$$

$$= \mathrm{tr}(\mathbf{A}_i|\phi_j\rangle\langle\phi_j|)$$

where

$$\mathbf{A}_i = \sum_{x \in A_i} |x\rangle\langle x|$$

is the projector of the event $x \in A_i$. In general, the pure state vectors can equivalently be written using the density matrices

$$\rho_0 = |\phi_0\rangle\langle\phi_0| \qquad \rho_1 = |\phi_1\rangle\langle\phi_1|$$

When the states are density matrices which might be mixed, we have that

$$p(\phi_i|\phi_j) = \mathrm{tr}(\mathbf{A}_i\rho_j)$$

The probability of a correct decision is then obtained when $\phi_i = \phi_j$ and weighting the original states by the prior probabilities q_is as follows:

$$Q_d = \sum_i q_i \text{tr}(\mathbf{A}_i \rho_i)$$

where the sum is computed over all the states and $\sum_i q_i = 1$. The probability of error is defined as

$$Q_e = 1 - Q_d$$

When there are two states, $p(\phi_1|\phi_1)$ is called probability of detection, and $p(\phi_1|\phi_0)$ is called probability of false alarm or fallout.

2.7.2   Optimal Detection

The general problem of detection is to define the subsets of observable values corresponding to the states in which the signal can be sent through the channel and to minimize the probability of error or maximize the probability of correct decision. In particular, given two states ϕ_0 and ϕ_1, the problem is to define one subset of observable values corresponding to, say, ϕ_1; the other subset of observable values is the complement; these subsets are often called "regions," the subset corresponding to ϕ_1 is called "region of acceptance," and the subset corresponding to ϕ_0 is called "region of rejection."

Using an algebraic description, the problem is to define the projectors of a vector space corresponding to the states. As the subset of observable values form a partition of the set of values, the collection of projectors form a resolution to unity of the vector space; for example, if two subsets of observable values have to be defined, two projectors $\mathbf{Q}_0$ and $\mathbf{Q}_1$ such that

$$\mathbf{Q}_0 + \mathbf{Q}_1 = 1 \qquad \mathbf{Q}_0 \mathbf{Q}_1 = 0$$

have to be defined for ϕ_0 and ϕ_1, respectively. Suppose the states are represented by two density matrices ρ_0 and ρ_1, respectively. The probability of correct decision is

$$\begin{aligned}
Q_d &= q_0 \text{tr}(\mathbf{Q}_0 \rho_0) + q_1 \text{tr}(\mathbf{Q}_1 \rho_1) \\
&= q_0 \text{tr}((1 - \mathbf{Q}_1) \rho_0) + q_1 \text{tr}(\mathbf{Q}_1 \rho_1) \\
&= q_0 \text{tr}(1 \rho_0 - \mathbf{Q}_1 \rho_0) + q_1 \text{tr}(\mathbf{Q}_1 \rho_1) \\
&= q_0 (\text{tr}(1 \rho_0) - \text{tr}(\mathbf{Q}_1 \rho_0)) + q_1 \text{tr}(\mathbf{Q}_1 \rho_1) \\
&= q_0 \text{tr}(1 \rho_0) - q_0 \text{tr}(\mathbf{Q}_1 \rho_0) + q_1 \text{tr}(\mathbf{Q}_1 \rho_1)
\end{aligned}$$

$$= q_0 \text{tr}(\rho_0) - q_0 \text{tr}(\mathbf{Q}_1 \rho_0) + q_1 \text{tr}(\mathbf{Q}_1 \rho_1)$$
$$= q_0 - q_0 \text{tr}(\mathbf{Q}_1 \rho_0) + q_1 \text{tr}(\mathbf{Q}_1 \rho_1)$$
$$= q_0 + \text{tr}((q_1 \rho_1 - q_0 \rho_0)\mathbf{Q}_1)$$

Using the same procedure, it is found that

$$Q_e = 1 - Q_d$$

It follows that the projectors being sought have to solve the following maximization problem:

$$\max_{\mathbf{Q}_1} \text{tr}((q_1 \rho_1 - q_0 \rho_0)\mathbf{Q}_1)$$

To this end, consider the SVD of $q_1 \rho_1 - q_0 \rho_0$, that is,

$$q_1 \rho_1 - q_0 \rho_0 = \sum_k \eta_k |\eta_k\rangle\langle\eta_k|$$

where the η_ks are the eigenvalues associated with the eigenvectors $|\eta_k\rangle$, and consider the probability of a correct decision expressed using the eigenvectors and the eigenvalues found by this decomposition:

$$Q_d = q_0 + \text{tr}((q_1 \rho_1 - q_0 \rho_0)\mathbf{Q}_1)$$
$$= q_0 + \text{tr}\left(\sum_k \eta_k |\eta_k\rangle\langle\eta_k|\mathbf{Q}_1\right)$$
$$= q_0 + \sum_k \text{tr}(\eta_k |\eta_k\rangle\langle\eta_k|\mathbf{Q}_1)$$
$$= q_0 + \sum_k \eta_k \text{tr}(|\eta_k\rangle\langle\eta_k|\mathbf{Q}_1)$$
$$= q_0 + \sum_k \eta_k \langle\eta_k|\mathbf{Q}_1|\eta_k\rangle \tag{2.11}$$

The crucial argument to find the optimal projectors that are the solution to this maximization problem is that

$$0 \le \langle\eta_k|\mathbf{Q}_1|\eta_k\rangle \le 1 \qquad \text{for all } k$$

since the eigenvectors are unit vectors and $\mathbf{Q}_1$ is Hermitian and with trace 1. Under these constraints, the maximum of (2.11) is obtained when

$$\eta_k > 0 \qquad \langle \eta_k | \mathbf{Q}_1 | \eta_k \rangle = 1$$

The second term of the maximizing argument is obtained if and only if

$$\mathbf{Q}_1 = |\eta_k\rangle\langle\eta_k|$$

It follows that the solution to the maximization problem is

$$\mathbf{Q}_1 = \sum_{\eta_k > 0} |\eta_k\rangle\langle\eta_k|$$

When the latter equality holds, we have that

$$Q_d = q_0 + \sum_{\eta_k > 0} \eta_k$$

2.7.3 Detection of Pure States

Consider two states described by two pure state vectors $|\phi_0\rangle$ and $|\phi_1\rangle$. It follows that

$$\rho_0 = |\phi_0\rangle\langle\phi_0| \qquad \rho_1 = |\phi_1\rangle\langle\phi_1|$$

It can be shown that the optimal projectors are the eigenvectors η_0 and η_1 of

$$q_1\rho_1 - q_0\rho_0 \tag{2.12}$$

and that the probabilities of correct decision and of error are

$$Q_d = \frac{1}{2}\left(1 + \sqrt{1 - 4q_0q_1|\langle\phi_0|\phi_1\rangle|^2}\right) \qquad Q_e = \frac{1}{2}\left(1 - \sqrt{1 - 4q_0q_1|\langle\phi_0|\phi_1\rangle|^2}\right)$$

where $|\gamma|^2 = |\langle\phi_0|\phi_1\rangle|^2$ is a measure of the distance between the pure state vectors; as for the details, see Helstrom (1976) and Cariolaro (2015).

The optimal projectors are what a detector has to observe in a particle in order to maximize the probability of correct decision. The optimal projectors are different from the pure state vectors; they are indeed a combination. Indeed, it can be shown that the optimal eigenvectors $|\eta_0\rangle$ and $|\eta_1\rangle$ are a linear combination of $|\phi_0\rangle$ and $|\phi_1\rangle$.

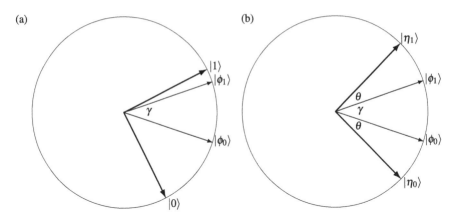

Fig. 2.20 Geometry of decision and eigenvectors. (**a**) Non-eigenvectors are mutually orthogonal but *asymmetrically* placed around $|\phi_0\rangle$ and $|\phi_1\rangle$. (**b**) The eigenvectors are mutually orthogonal and *symmetrically* placed around $|\phi_0\rangle$ and $|\phi_1\rangle$

If the pure state vectors represent the superposition state of a qubit, the optimal eigenvectors represent how a device should be set in order to optimally detect the state of the qubits; for example, if the pure state vectors represent a polarized direction (e.g., vertical), the optimal eigenvectors represent how a PBS has to be angled, the angle being that between the eigenvectors and the state vectors.

As a result, using $|\eta_0\rangle$ and $|\eta_1\rangle$ is like "cutting" the space of observable values in an "oblique" way to the way in which the space is cut by using classical observable vectors like $|0\rangle$ and $|1\rangle$. The probability of detection and the probability of false alarm are given by the angle between the pure state vectors $|\phi_0\rangle$, $|\phi_1\rangle$ and by the angles between the eigenvectors and $|\phi_0\rangle$, $|\phi_1\rangle$. Therefore, $|\langle\phi_0|\phi_1\rangle|^2$ determines the geometry of the decision between the two states. Figure 2.20a illustrates the geometric interpretation. If one looks for the eigenvectors, one can see that

$$\theta = \frac{\frac{\pi}{2} - \gamma}{2}$$

so that the eigenvectors are "symmetrically" located around the state vectors. The replacement of η_0 and η_1 with θ yields the minimal probability of error.

To obtain this minimum, a detection device should be set as follows. The region of acceptance is defined, assuming that the density matrices are mixed. Defining the region of acceptance means that the knob of the detection device is set to a value of x, say, 1, as depicted in Fig. 2.21a; if the device measures 1, it decides a_1; otherwise, it decides a_0. Then, the angles are calculated and the knob is rotated by the angle between $|1\rangle$ and $|\phi_1\rangle$ as depicted in Fig. 2.21b; if the device measures η_1, it decides a_1; otherwise, it decides a_0.

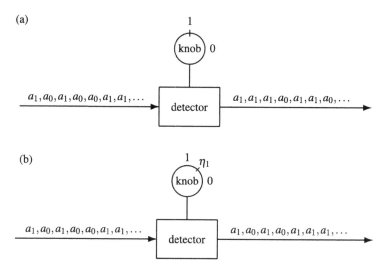

Fig. 2.21 Optimal detection and knob rotation

2.8 Suggested Readings

Kumar (2008) wrote an introduction to the history of QM; another historical introduction was written by Renn (2006).[3] Zeilinger (2010) is an enlightening introduction to superposition, interference, and entanglement. Another book of the same kind was published by Lederman and Hill (2011) as an introduction to quantum physics. Nielsen and Chuang (2000) is one of the most exhaustive books of quantum computing, while Rieffel and Polak (2011) provides a clean mathematical explanation.

The book written by Hughes (1989) and the first part of the book of Griffiths (2002) give an exhaustive illustration of QM where the notion of observable is specifically addressed together with the issues of incompatibility. The latter was addressed by van Rijsbergen (2004) by an example of incompatibility between relevance and aboutness. Peres (2002) is an introduction to quantum physics.

The first introduction to superposition was provided by Dirac (1935) who also introduced the bra(c)ket or Dirac notation. Albert (1994) is another clear introduction to superposition and other concepts of QM. Jauch (1968) provides a perspective from the logic's point of view. Qubits are, for example, explained by Mermin (2007), Nielsen and Chuang (2000), and Yanofsky and Mannucci (2008), while Aaronson (2013) provides, among other things, the intuition based on the additional dimension for explaining the need of complex numbers. Meglicki (2008) illustrates a concrete description of qubit and quantum gate implementation, which

---

[3]I read the translation in Italian.

is also described by Yanofsky and Mannucci (2008) for computer scientists; the latter also gives an explanation of the Bloch sphere.

A full account of Gleason (1957)'s theorem is provided by Hughes (1989), while its importance in IR was pointed out by van Rijsbergen (2004). Wootters (1981) introduced the statistical distance as the reason that the squared cosine of angle is the probability function of a Bernoulli trial and has inspired the explanation of the square rule or in general the trace rule. The reason that transformations of probability are linear has been inspired by Hardy (2001). The proof of the no-cloning theorem is provided by Rieffel and Polak (2011). The illustration of conditional probability in the quantum mechanical framework and an explanation of why the trace rule is the rule for computing probabilities are derived from Aaronson (2013)'s book, which is also worth reading for an unconventional view of QM.

Interference and the double-slit experiment were illustrated by Feynman et al. (1965); many other authors explain these concepts such as Hughes (1989). Accardi (1984) and Accardi (1997) suggested some complementary considerations on the validity of Feynman's explanations of the "strange" probability distributions. Another illustration was given by Fine (1973).

Entanglement has been illustrated by, for example, Zeilinger (2010) who gives a detailed account of the experiments permitting to observe entanglement and by Nielsen and Chuang (2000) and Rieffel and Polak (2011) who give a mathematical and computational account of this phenomenon and explain how Bell's inequality can be obtained and the conditions in which it might be violated. The complete description of the Bell inequalities was provided by Bell (1964, 1987) by replying to the comments by Einstein et al. (1935). Other inequalities were proposed and investigated by Accardi and Fedullo (1982) as for the role played by the Hilbert spaces and by Pitowsky (1989) with an emphasis on logic and statistics.

The main source for quantum detection theory was written by Helstrom (1976). Cariolaro (2015) is a beautiful book on quantum communications from a signal and communication theory perspective; it provides a clear introduction to QM and other related topics. An illustration of the statistical science behind detection and in general quantum statistics was provided by Barndorff-Nielsen et al. (2003) and Malley and Hornstein (1993).

Chapter 3
Quantum Mechanics and Information Retrieval

The first two chapters have introduced the elements of IR and QM with the aim of describing the notions intersected by both disciplines. In this chapter, we describe how this intersection has been implemented. We selected and presented in no predefined order the most significant contributions to the implementation of this intersection; some contributions appear to be less mature than others; however, we decided to include them since they are sources of future work. Other contributions might appear less "quantum inspired" than others; however, each research work contains concepts and tools that are somehow linked to the quantum mechanical framework illustrated in the book. In the end, the contributions reported in this chapter cover a wide range of issues, from modeling issues to user interaction issues. The chapter ends with suggestions of further reading.

3.1 Introduction

The implementation of the intersection between QM and IR illustrated in this chapter has been based on the connections between the mathematical framework used both in QM and IR; a glaring example of these connections is given by the connection between VSM and the Hilbert spaces where the former can be viewed as a simple application of the latter. This implementation allowed the researchers in IR to adopt powerful means of expression and generalization of their models in an attempt to achieve better retrieval performance. Moreover, the connections between the mathematical framework used both in QM and IR aimed at investigating the existence of quantum phenomena (e.g., superposition, interference, entanglement) which may be at the basis of the problems addressed in IR, such as the difficulty caused by incompatibility in capturing relevance through the aboutness of a document given a query.

© Springer-Verlag Berlin Heidelberg 2015   101
M. Melucci, *Introduction to Information Retrieval and Quantum Mechanics*,
The Information Retrieval Series 35, DOI 10.1007/978-3-662-48313-8_3

This chapter illustrates different implementations of the intersection between QM and IR. These implementations have been illustrated in this chapter in no preferred order if not the chronological order suggested by the relevant literature published since the first decade of this century. This chapter is organized in the following sections. Section 3.2 briefly presents the motivations and the applications of the use of the formalism of the quantum mechanical framework in IR. Section 3.3 concentrates on kinds which are the logical construct used for introducing a logic following quantum principles. Section 3.4 reviews some contributions to the research in concept combination based on superposition and entanglement. Section 3.5 describes the research on word disambiguation and vector negation which are based on the theory of vector spaces. Section 3.6 reviews some contributions to the research on semantic spaces based on entanglement. Section 3.7 describes how context can be modeled using the theory of vector spaces. Section 3.8 illustrates a PRP based on the idea that documents are not inspected by the user one at a time, but that the user can find himself in a superposition state. Section 3.9 briefly describes how implicit feedback, polyrepresentation, and information need representation were addressed using the quantum mechanical framework. Section 3.10 describes an application of quantum detection to IR and in particular how an alternative PRP can be defined. Section 3.11 reports on some experimental investigations. Finally, Sect. 3.12 concludes with some suggestions for further reading.

3.2   Quantum Formalism

The book *The Geometry of IR (GIR)* by van Rijsbergen (2004) was the first contribution to the intersection between IR and QM. The main aim of the book was to give insights into how probability, logic, and vector spaces can be combined into the formalism of QM. This combination was driven by the need to describe in a formal way how users interact with a system. The need of a formal way can be explained by two main yet related motivations, that is, that without a formal way, the derivation of methods computable by systems is impossible or at least very difficult and it is also impossible or at least very difficult to make predictions about the systems. The author of GIR paralleled von Neumann (1955) who described quantum physics using the mathematical concepts of logics, geometry, and probability in a consistent way but without an explicit reference to the physical world to which QM was referred by physicists. The connection between the GIR and the theoretical approach to QM of von Neumann was not only at the descriptive level though. This connection was also motivated by the analogy between what happens when subatomic particles are observed and what happens when a user interacts with a document. When a user interacts with a document, she is ignoring that the document has been indexed and that a document representation has been stored by the system. From this point of view, a document might not have any representation implemented within the indexes of a system.

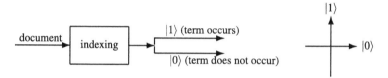

Fig. 3.1 Term occurrence, observables, and vector space basis

3.2.1 Observables and Projectors

Section 2.2 introduces the dimensionality d of the vector space as the number of distinct values taken by the observable; for example, the observable that describes the relevance of the information contained in a document with respect to an information need often takes $d = 2$ values (i.e., relevant or 1 and nonrelevant or 0) and the vector space, and then the basis is bidimensional. It follows that term occurrence corresponds to a basis of two vectors as depicted in Fig. 3.1 and formalized as follows:

$$
\begin{array}{ccc}
\text{Observable outcome} & \text{Value} & \text{Ket} \\
\text{term occurs} & 1 & |1\rangle = \begin{pmatrix} 1 \\ 0 \end{pmatrix} \\
\text{term does not occur} & 0 & |0\rangle = \begin{pmatrix} 0 \\ 1 \end{pmatrix}
\end{array}
$$

The second column of the table includes the values taken by a variable used in IR to represent term occurrence. Other labels would be used to define other events; for example, the basis vector denoting the event that positive numbers are observed can be written as $|+\rangle$, whereas the one denoting the event that negative numbers are observed can be written as $|-\rangle$.

Note that the basis vectors of this observable are mutually orthogonal, thus meaning that the events (e.g., "term occurs" and "term does not occur") are mutually exclusive; for example, two orthogonal basis vectors corresponding to two mutually exclusive events may be

$$
|1\rangle = \begin{pmatrix} 1 \\ 0 \end{pmatrix} \qquad |0\rangle = \begin{pmatrix} 0 \\ 1 \end{pmatrix}
$$

and they can exist in the same vector space of

$$
|+\rangle = \begin{pmatrix} \frac{1}{\sqrt{2}} \\ \frac{1}{\sqrt{2}} \end{pmatrix} \qquad |-\rangle = \begin{pmatrix} \frac{1}{\sqrt{2}} \\ -\frac{1}{\sqrt{2}} \end{pmatrix}
$$

The extension to two or more dimensions follows the route followed in Sect. 2.2. Consider, for example, the following pairs of observables: the occurrence of two terms, term occurrence and relevance, aboutness and relevance, and document retention and term occurrence. If the occurrence of two terms is considered, the basis vector corresponding to the event that neither term occurs is

$$|00\rangle = \begin{pmatrix} 0 \\ 0 \\ 0 \\ 1 \end{pmatrix}$$

This vector is obtained by the tensor product of

$$|0\rangle = \begin{pmatrix} 0 \\ 1 \end{pmatrix}$$

for both observables.

3.2.2   Indexing and Retrieval as Measurement

Following the quantum mechanical framework, describing document indexing and retrieval in terms of measurement is possible, that is, the representation of the document is the result of the measurement performed when the user is interacting with the document. Along such a parallel, what is observed during the interaction between the user and the document is the outcome of a measurement, and it is considered as a representation of the document; for example, the click-through data that result from the interaction between the user and a search engine result page (serp) may be viewed as the outcome of a measurement, and it is considered as a representation of the document; these data may be the click itself, the portion of serp looked at by the user's eyes, and the portion of page on which the user clicked; in other words, it is not necessarily the bag of words extracted for representing the document informative content.

Moreover, the document that is subjected to the interaction with the user can simply be represented as a vector or more precisely by a state vector; it is labeled "state" because it is representing the state of the object (e.g., the document) subjected to measurement. However, a state vector is more than a simple vector: it is indeed a "container" of all the possible outcomes of all the possible measurements. This is to say that whatever the measurement is, there exists one state vector which provides the distribution of probability of the outcomes of every possible measurement. In IR terms, whatever the query or any other interaction with the document is, there exists one state vector which provides the distribution of probability of relevance, aboutness, or any other measurement. Therefore, the distribution of probability of, say, relevance of a document is not provided by an index but by the state vector of the document.

3.2.3   State Vector and Vector Space Model

A state vector is not exactly a vector of the VSM. In the VSM, a document is a vector because it is a linear combination of basis vectors which correspond to the index terms (see Sect. 1.3). As the index terms are finite, the dimensionality of the vector space is finite. Moreover, there is no notion of distribution of probability in the VSM. Finally, the vectors of the VSM are defined over the real field. In contrast, the state vector that represents a document in the GIR is not necessarily a linear combination of other vectors corresponding to something else. A state vector may lie in an infinite vector space since it is "only" an abstract mathematical object. Moreover, a state vector induces a number of distributions of probability stemming from a number of possible measurements to which the document may be subjected. Finally, the state vector of QM is defined over the complex field; the reasons are not only confined to the will of enlarging the representational power of the state vector or to examples such as the TFIDF weighting scheme, which is composed of two real numbers as a complex number is. The complex field is necessary because the quantum mechanical framework requires the use of the complex field for finding the solutions to the mathematical problems of the framework as explained in Sect. A.4.

A state vector is an abstract mathematical object defined over an infinite space, thus meaning that it incorporates infinite outcomes of a measurement. This infiniteness may not be realistic in IR where measurements are usually finite (e.g., term frequency or relevance). However, there is another infiniteness which is significant to the purposes of using a state vector to represent a document. A state vector can simultaneously be the linear combination of infinite possible measurements in the same way a vector can simultaneously be the linear combination of infinite possible bases, the latter being a fact of linear algebra stating that a vector can be expressed in many different ways by using different bases and different coordinates; for example:

$$\frac{1}{\sqrt{5}}\begin{pmatrix} 1 \\ 2 \end{pmatrix} = \frac{1}{\sqrt{5}}\begin{pmatrix} 1 \\ 0 \end{pmatrix} + \frac{2}{\sqrt{5}}\begin{pmatrix} 0 \\ 1 \end{pmatrix} \tag{3.1}$$

$$= \left(\frac{1}{\sqrt{2}\sqrt{5}} + 1 \right) \begin{pmatrix} \frac{1}{\sqrt{2}} \\ \frac{1}{\sqrt{2}} \end{pmatrix} + \left(\frac{1}{\sqrt{2}\sqrt{5}} - 1 \right) \begin{pmatrix} \frac{1}{\sqrt{2}} \\ -\frac{1}{\sqrt{2}} \end{pmatrix} \tag{3.2}$$

where (3.1) may refer to the occurrence of a query term and (3.2) may refer to relevance. In particular, the basis vectors

$$\begin{pmatrix} 1 \\ 0 \end{pmatrix} \quad \begin{pmatrix} 0 \\ 1 \end{pmatrix}$$

implement the observable of query term occurrence, and the basis vectors

$$\begin{pmatrix} \frac{1}{\sqrt{2}} \\ \frac{1}{\sqrt{2}} \end{pmatrix} \quad \begin{pmatrix} \frac{1}{\sqrt{2}} \\ -\frac{1}{\sqrt{2}} \end{pmatrix}$$

implement the observable of relevance. Both (3.1) and (3.2) define a subspace of the vector space spanned by the document state vector. Therefore, the subspace spanned by a basis vector is contained by the subspace spanned by the document state vector, but the subspace spanned by basis vector of (3.1) is not contained by any subspace spanned by the basis vectors of (3.2).

As a state vector is in general defined over the complex field, the coordinates of (3.1) and (3.2) may in general be complex, thus distinguishing the notion of state vector from the notion of document vector of the VSM. These coordinates are the result of the projection of state vector on the subspace spanned by the respective basis vector, but the result of a projection is not necessarily measuring a size, and then it is not a real number. The trace rule, which becomes the square rule when only vectors are used, dictates that the modulus of a coordinate is between 0 and 1, and the squared modulus is the probability that the measurement of an observable (e.g., query term occurrence) on the document state vector yields true, that is, the probability that the query term occurs in the document represented by the state vector.

3.2.4   Probability, Logic and Geometry

Suppose that the projectors

$$\begin{pmatrix} 1 & 0 \\ 0 & 0 \end{pmatrix} \quad \begin{pmatrix} 0 & 0 \\ 0 & 1 \end{pmatrix}$$

represent the answers to a two-valued question such as "does a query term occur?". The geometry of the space is exploited for calculating the probability of a particular answer. The document state vector has to be projected orthogonally down onto the subspace of the answer (i.e., the subspace spanned by the projector of the answer), and the size of that projection has to be measured using the trace rule for obtaining the probability. Probability measures have to sum to 1 since the subspaces are mutually orthogonal. As the projectors of the answers are one dimensional, they

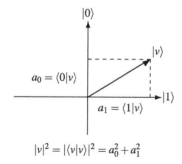

Fig. 3.2 Pythagoras' theorem

span a ray, and therefore, they correspond to a vector (i.e., the vector which spans the ray). As these projectors are mutually orthogonal, the corresponding vectors are mutually orthogonal too. It follows that the square rule corresponds to Pythagoras' theorem as depicted in Fig. 3.2. In this way, the geometry of the space provides the rules for calculating the probabilities, thus establishing the connection between geometry and probability. The reasons that the square rule and in general the trace rule are used for calculating the probability of a particular answer are illustrated in Sect. A.2.

3.2.5 Gleason's Theorem

The use of subspaces, trace rule, and operators on projectors is however insufficient for explaining why they are different sides of the same coin. Thus, the question is therefore how to connect geometry (subspaces), probability (trace rule), and logic (projectors) in a way that they *have to* "live" together. van Rijsbergen (2004) refers to the answer contained in the Gleason theorem which states that if we have an observable, the possible answers represented as projectors, and the probabilities of these answers provided by some external sources, then the density matrix resulting from the linear combination of these projectors is the one that can reproduce these probabilities (see also Sect. A.2). This theorem is a sort of "comfort theorem" ensuring that if a state assigns to some projectors the corresponding probability values, then we can represent those values through an algebraic calculation, i.e., the trace rule applied on the density matrix given by the theorem.

The Gleason theorem integrates three main modeling approaches in IR (i.e., Boolean, vector space, and probabilistic) within a common framework provided by the theory of complex vector spaces and the probability theory thereof as depicted in Fig. 3.3.

The Boolean modeling refers to the use of the set theory and classical logic to represent how the informative content of documents is represented and how a system decides whether to retrieve a document for answering a user's query (Sect. 1.2). The basic idea of the Boolean modeling is that a term (e.g., a keyword) is a set

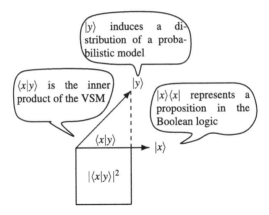

Fig. 3.3 Gleason's theorem in Information Retrieval

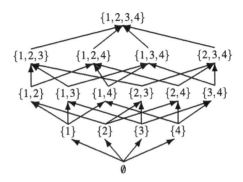

Fig. 3.4 Boolean lattice

of documents (i.e., the documents indexed by that term) and that a query is a proposition which is either true or false according to the document to which the query is applied. The subsets of documents induced by a vocabulary of terms can be organized through inclusion relationships in order to create a conceptual structure called Boolean lattice; an example is depicted in Fig. 3.4 where each node of the lattice corresponds to a subset and each oriented edge corresponds to an inclusion relationship (e.g., $\{1, 3\}$ is a subset of $\{1, 2, 3\}$ and $\{1, 3, 4\}$). Each subset of the lattice is a one-to-one correspondence with a projector and therefore a subspace of a complex vector space; for example, the subset of documents indexed by keyword 1 but not by keyword 2 can correspond to the basis vector $|10\rangle$ and then to the rank-one projector $|10\rangle\langle10|$. These projectors are those mentioned in the Gleason theorem.

The VSM provides that the terms of a document collection are represented by basis vectors, documents are represented as vectors, and queries are represented as state vectors. An IR system decides whether to retrieve a document according to the value of the inner product between a query vector and the document vector (see also Sect. 1.3). When both vectors are unit vectors (i.e., they have length or norm 1), the inner product can become a special case of the trace rule; for example, a query y may be represented by $|y\rangle$, a document x may be represented by $|x\rangle$ such that $|\langle y|y\rangle|^2 = |\langle x|x\rangle|^2 = 1$, and a system can rank the documents by $|\langle x|x\rangle|^2$. It follows that the inner product used in the VSM is a special case of the trace rule, a query vector is a special case of state, and a document is an example of projector mentioned in the Gleason theorem.

Probabilistic modeling can fit in the Gleason theorem because it is based on events and probability measures where the events can be represented by projectors and the probability values assigned to the projectors come from the probability function used in the given probabilistic model; for example, the probability values are provided by the Bernoullian function of the BIR model and are assigned to the projectors corresponding to the string of binary digits representing term occurrence; the Gleason theorem assures that the probability function of the BIR model is a state and that this state can only be implemented as a density matrix.

3.2.6   Incompatibility and Relevance

Incompatibility arises when two observables cannot be simultaneously measured; otherwise, the outcome of one observable would disturb the other observable which could not be measured with arbitrary precision. The notion of incompatibility can hardly be translated into an IR situation; for example, when measuring the relevance of a document, the mind of a user should be viewed as a particle able to store the state resulting from the assessment of the relevance of two documents as depicted in Fig. 3.5. At the beginning, a user is presented with two documents A and B and is asked to assess the relevance of A. After this assessment, the user stores the assessment in his mind as in Fig. 3.5a. He is then presented with document B and asked to assess the relevance of B. The informative content of B can however

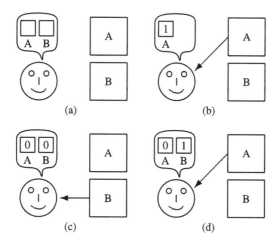

Fig. 3.5 Incompatibility of the measurement of relevance

influence the assessment of the relevance of A stored in the user's mind, thus changing the latter as depicted in Fig. 3.5c; indeed, the content of B might be more authoritative than the content of A, and the user might prefer B and no longer prefer A when, for example, he has time enough to read one document only. If the user is again presented with A, he may change or confirm the assessment of the relevance of A, but the assessment of the relevance of B might be disturbed as in Fig. 3.5d. However, incompatible observables are not the only means to represent the dependency between two document relevant assessments. Using conditional probability is another means: suppose A is relevant, i.e., P(A is relevant) = 1; suppose B is also relevant; it follows that P(A is relevant|B is relevant) may be different from P(A is relevant).

The difference between incompatibility and conditionality is that the former operates on observables, whereas the latter operates on probability, that is, incompatibility changes the range of observable values, whereas conditionality changes the probability of observable values that cannot change. Suppose, for example, a document is a priori relevant with a certain probability. After observing an index term in the document, the probability of relevance of the document can a posteriori be changed by the likelihood and Bayes' rule. Although the distribution of probability has changed, the observable values are the same. In contrast, suppose the observation of the index term changes the value of the "relevance" observable and moves the state of the document to a superposition state corresponding to a relevance degree not previously observed. The probability of the latter relevance degree can only be computed by the rules of the quantum mechanical framework, and not only the distribution of probability has changed, the range of observable values has also changed.

van Rijsbergen (2004) illustrated an example of incompatibility between two observables using relevance and aboutness, the latter often being considered distinct from the former since relevance refers to a user's information need, whereas aboutness refers to a topic. Aboutness answers the question "is this document about this topic?", whereas relevance answers the question "is this document relevant for this information need?". He assumes an incompatibility between relevance and aboutness such that when relevance is observed and a relevance value (e.g., "this document is relevant") is obtained, the outcome of the measurement of aboutness is inevitably imprecise, and this outcome might be different from the outcome that would be observed if aboutness were measured before measuring relevance.

It should be noted that the decision of representing the conjoint measurement of two observables using operators acting on vector spaces derives from the assumption that incompatibility is actually observed in the physical world and that it can be modeled using vector spaces. Indeed, if incompatibility were not observed, the conjoint measurement of two observables could be represented by a collection of four projectors defined in the quadridimensional space as explained in Sect. 2.2. It is the occurrence of incompatibility that induces the definition of two appropriate noncommutative operators and of an appropriate state vector. This is to say that the algebra of projectors and density matrix is only a language for describing what is observed in the real world, and it is not the real world.

3.2.7 Entanglement and Correlation

In Sect. 2.4, it was explained that the classical uncertainty addressed in probability and statistics is about the observables measured in ensembles such as collections of documents indexed by terms; the occurrence of a term in a document drawn from a collection is an uncertain event because the collection may contain both documents indexed by the term and documents not indexed by the term.

In the quantum mechanical framework, uncertainty may occur not only when the elements are collected in an ensemble but also when each of them is in a superposed state (e.g., the uncolored balls mentioned in Sect. 2.4) since the color is observed when it is measured and the state of the ball randomly collapses to the color asked by the measurement with probability given by the probability amplitudes of the superposition.

Corresponding to these two types of uncertainty, there are two main types of density matrix, i.e., pure distributions which represent the uncertainty of superposed states and mixed distributions which represent the uncertainty of ensembles. Mathematically, pure distributions correspond to rank-one projectors (i.e., pure density matrices), and mixed distributions correspond to mixtures of projectors such that the mixture weights are the probabilities of the events represented by the projectors. A density matrix associated with two mutually exclusive binary events is a mixed distribution when it is a mixture of four orthogonal projectors in a four-dimensional Hilbert space.

In IR, correlation has been often utilized to implement various methods; for example, query expansion and relevance feedback are somehow based on term correlation, whereas document clustering is based on document correlation. The basic idea of correlation is that an observable can help predict another observable; for example, the occurrence of a term can help predict the relevance of a document containing the term, the latter correlation being at the basis of the cluster hypothesis in IR.

In QM, correlation has been also an important topic; it is at the basis of crucial applications such as cryptography and of visionary uses such as teleportation as explained by Nielsen and Chuang (2000). Quantum correlation presents some peculiarities which led to the use of the word "entanglement." Entanglement is a sort of correlation between the observables measured in atomic-size particles such as photons when these particles are not necessarily collected in ensembles.

In IR, the situation that might be similar to that encountered in physics might consist of a single document in which two observables (e.g., relevance and aboutness) are measured without counting on an ensemble of documents already labeled in terms of these two observables. Despite entanglement being a kind of correlation, there are some basic differences between entanglement and the classical correlation encountered in the macroscopic world such as the world of IR.

Suppose the elements of an ensemble are described by two observables with binary outcome (e.g., either 0 or 1) so that each element can be in one state given by the possible combination of the binary outcomes (e.g., $|00\rangle$, $|01\rangle$, $|10\rangle$, $|11\rangle$); for example, the document of a collection can be described by the occurrence of two terms. It follows that

$$\mathbf{C}_{ij} = |ij\rangle\langle ij| \qquad i = 0, 1 \quad j = 0, 1$$

In general, a mixed distribution combines distributions. When there are four possible states of the elements of an ensemble, the mixed distribution can be as follows:

$$\rho = p_{00}\mathbf{C}_{00} + p_{01}\mathbf{C}_{01} + p_{10}\mathbf{C}_{10} + p_{11}\mathbf{C}_{11} \qquad (3.3)$$

The basic idea underlying ρ is based on the following correspondence: $\mathbf{C}$ represents a state, while p represents the probability that an element of the ensemble is in that state. The density matrix (3.3) is called "separable" since the states can be written as two separated states tensored together as follows:

$$|ij\rangle\langle ij| = |i\rangle\langle i| \otimes |j\rangle\langle j| \qquad i = 0, 1 \quad j = 0, 1$$

Separability enables the expression of the event represented by $\mathbf{C}_{ij}$ as the product of two distinct events; thus, it allows us to express the fact that a four-event set is the product of two binary event sets, that is, $\{00, 01, 10, 11\}$ where the first bit refers to the first event set and the second bit refers to the second event set; for example, $|11\rangle\langle 11|$ represents the event that two terms occur in a document.

When (3.3) represents a separable state, all the correlation comes from the correlation between the two observables (e.g., the occurrence of one term and the occurrence of the other term) and can be measured in the probability distribution $p_{00}, p_{01}, p_{10}, p_{11}$ estimated from sampling an ensemble (e.g., a document collection). When this distribution is uncorrelated, the density matrix (3.3) is called "uncorrelated" and can be written as

$$\rho_A \otimes \rho_B$$

where ρ_A is the density matrix of the distribution of, say, the occurrence of one term and ρ_B is the density matrix of the distribution of the occurrence of the other term. In this way, the probability that the occurrence is ij is the product of the marginal probability that the occurrence of the first term is i and the marginal probability that the occurrence of the second term is j, that is, the occurrences are statistically independent.

Consider the source of correlation stemmed when an element is in a superposed state. As all the uncertainty about an observable comes from it being in a superposed state, all the correlation between two observables measured in this element comes from it being in a superposed state. As the outcome of one observable is uncertain because the state of the element collapses to an observable value in a random manner, the correlation between two observables comes from the random behavior of the pairs of collapses to the possible outcomes of the measurement in the element. Considering an ensemble is unnecessary since a single element in a superposed state is sufficient to measure the correlation between two observables as described in Sect. 2.6 with regard to pairs of entangled photons. Note that when an element is not in a superposed state and it is on the contrary in a ground state (e.g., either a document contains a term or not), the measurement of two observables (e.g., term occurrences) can result in the ground state or not since such a state behaves like an event which can be either true or false; in this case, no uncertainty nor correlation can be observed. Consider (3.3) and suppose the state ρ of an element is not mixed so that all the uncertainty comes from it being a superposition $|\phi\rangle$. It follows that

$$\rho = |\phi\rangle\langle\phi|$$

This state is separable when the superposed state can be expressed as tensor product between the states corresponding to what is measured by two observables. Similarly to $|ij\rangle$, $|\phi\rangle$ is separable when

$$|\phi\rangle = |\phi_A\rangle \otimes |\phi_B\rangle$$

which implies that $\rho = |\phi_A\rangle\langle\phi_A|\otimes|\phi_B\rangle\langle\phi_B|$. When ρ is not separable, it is *entangled*, and all the correlation between two observables comes from it being entangled even if it is the state of the only element of an ensemble.

While the measure of correlation is provided by the mixture weights (i.e., probabilities), a measure of entanglement can only be provided by the projectors of the mixture. A measure of entanglement, i.e., how much entanglement occurs in a density matrix, can be provided by the Schmidt number, that is, the number of nonzero eigenvalues of the Schmidt decomposition (see Sect. A.8).

3.3 Using Kinds

The GIR introduced in Sect. 3.2 is not only about probability and vector spaces, it is also about logic. The emphasis on logic was due to the aim of finding an alternative retrieval language to the retrieval languages based on subsets, that is, the classical Boolean logic, and of "investigating a semantics based on subspaces in a Hilbert space and see what kind of retrieval language corresponds to it" as argued by van Rijsbergen (2004). A possible implementation of the logic induced by subspaces might be based on kinds.

3.3.1 Definition of Kind

A concept may be described by its intension (i.e., the set of traits that characterize the concept) or by its extension (i.e., the set of individuals of the concept). A kind is the twofold representation of a concept—the set of individuals of the concept on the one hand and its traits (or attributes) on the other; for example, two kinds might be $K_1 = (\{a,b,c\},\{x,y\})$ and $K_2 = (\{c,d\},\{y,z\})$ where a,b,c are individuals described by the traits x and y and both traits y and z describe both individuals c and d; examples of individuals are documents and examples of traits are index terms.

More formally, suppose T_* is a set of traits used for describing the individuals collected in A_*. Consider traits that represent individuals. A kind K is a pair (A, T) where A is a subset of individuals and T is a subset of traits; for example, if $A_* = \{a_1, \ldots, a_n\}$ is a set of n individuals (e.g., documents) and $T_* = \{t_1, \ldots, t_k\}$ is a set of k traits (e.g., index terms), a legitimate kind may be $K = (\{a_1, a_2, a_3\}, \{t_2, t_4\})$.

Two functions are defined on a kind given a set A of individuals and a set T of traits:

- For every A, the function $\mathrm{tr}(A)$ returns the subset of traits that describes every individual of A; for example,

$$\mathrm{tr}(\{a,b,c\}) = \{x,y\} \qquad \mathrm{tr}(\{c,d\}) = \{y,z\}$$

- For every T, the function $\mathrm{in}(T)$ returns the subset of individuals described by all the traits of T; for example,

$$\mathrm{in}(\{x,y\}) = \{a,b,c\} \qquad \mathrm{in}(\{y,z\}) = \{c,d\}$$

The definition and the examples of these two functions highlight that a kind is characterized by the fact that every individual in A instantiates every trait in T and no individual not in A instantiates every trait in T. It is possible to write that

$$A = \mathrm{in}(T) \qquad T = \mathrm{tr}(A)$$

A pictorial illustration is shown in Fig. 3.6 where the individuals correspond to the rows of the matrix and the traits correspond to the columns. When the element of row i and column j is 1, the individual i is described by the trait j. In this way, a kind is shaped as a rectangle of 1s.

1	1	1	0	0	0	1	1
1	1	1	0	0	0	1	1
1	1	1	0	0	0	1	1
0	1	1	1	0	0	1	1
0	1	1	1	0	0	1	1
0	1	1	1	0	0	1	1
0	1	1	1	0	0	1	1
0	1	1	1	1	1	1	1
0	1	1	1	1	1	1	1
0	0	0	1	1	1	0	0
0	0	0	1	1	1	0	0
0	0	0	1	1	1	0	0

Fig. 3.6 Some kinds

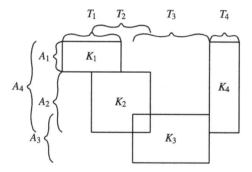

Fig. 3.7 Disjunction and conjunction of kinds

3.3.2 Operations on Kinds

Similarly to subsets which can be provided with disjunction and conjunction (i.e., intersection), the kinds can be provided with meet (disjunction) and join (conjunction) operations which are defined as follows and depicted in Fig. 3.7:

- Meet is defined as $K_1 \vee K_2 = (\text{in}(T_1 \cap T_2), T_1 \cap T_2)$.
- Join is defined as $K_1 \wedge K_2 = (A_1 \cap A_2, \text{tr}(A_1 \cap A_2))$.

It is possible to show that the meet and join of any pair of kinds are kinds themselves, thus allowing us to write expressions using kinds in a closed form. Indeed

$$\text{tr}(\text{in}(T_1 \cap T_2)) = T_1 \cap T_2 \qquad \text{in}(\text{tr}(A_1 \cap A_2)) = A_1 \cap A_2$$

A containment relation is defined as

$$K_1 \leq K_2$$

if and only if

$$K_1 = K_1 \wedge K_2$$

Similarly to subsets, there exists a minimum kind **0** and maximum kind **1** defined as follows:

$$\mathbf{1} = (A_*, \emptyset) \qquad \mathbf{0} = (\emptyset, T_*)$$

Moreover, it can be shown that

$$\mathbf{1} \geq K \qquad \mathbf{0} \leq K$$

for every kind K.

The operators used to define the minimum (or null) kind $\mathbf{0}$ and the maximum (or identity) kind $\mathbf{1}$ cannot be the same as those used for defining the empty set and its complement of a collection of subsets. The correctness of the definition of minimum kind $\mathbf{0}$ and maximum kind $\mathbf{1}$ can be seen as follows:

- Joining a kind and the empty kind results in the empty kind similarly to the intersection between a set and the empty set. Indeed

$$K \wedge \mathbf{0} = (A \cap \emptyset, \mathrm{tr}(A \cap \emptyset)) = (\emptyset, \mathrm{tr}(\emptyset)) = (\emptyset, T_*) = \mathbf{0}$$

- Joining a kind and the identity kind results in the kind similarly to the intersection between a set and the identity set. Indeed

$$K \wedge \mathbf{1} = (A \cap A_*, \mathrm{tr}(A \cap A_*)) = (A, \mathrm{tr}(A)) = K$$

- Meeting a kind with the empty kind results in the kind similarly to the union between a set and the empty set. Indeed

$$K \vee \mathbf{0} = (\mathrm{in}(T \cap T_*), T \cap T_*) = (\mathrm{in}(T), T) = (A, T) = K$$

- Meeting a kind with the identity kind results in the identity kind similarly to the union between a set and the identity set. Indeed

$$K \vee \mathbf{1} = (\mathrm{in}(T \cap \emptyset), T \cap \emptyset) = (\mathrm{in}(\emptyset), \emptyset) = (A_*, \emptyset) = \mathbf{1}$$

In particular, it follows that

$$\mathbf{0} \vee \mathbf{0} = (\mathrm{in}(T_* \cap T_*), T_* \cap T_*) = (\mathrm{in}(T_*), T_*) = (\emptyset, T_*) = \mathbf{0}$$

$$\mathbf{1} \vee \mathbf{1} = (\mathrm{in}(\emptyset \cap \emptyset), \emptyset \cap \emptyset) = (\mathrm{in}(\emptyset), \emptyset) = (A_*, \emptyset) = \mathbf{1}$$

$$\mathbf{0} \wedge \mathbf{0} = (\emptyset \cap \emptyset, \mathrm{tr}(\emptyset \cap \emptyset)) = (\emptyset, \mathrm{tr}(\emptyset)) = (\emptyset, T_*) = \mathbf{0}$$

$$\mathbf{1} \wedge \mathbf{1} = (A_* \cap A_*, \mathrm{tr}(A_* \cap A_*)) = (A_*, \mathrm{tr}(A_*)) = (A_*, \emptyset) = \mathbf{1}$$

All this mathematical formulation lies in the interest in kinds explained by the fact that the distributive law does not hold for them and their operators for disjunction and conjunction. To show that the distributive law does not hold when meeting and joining kinds, suppose three kinds K_1, K_2, K_3 can be chosen so that

$$K_1 \wedge (K_2 \vee K_3) \neq \mathbf{0}$$

and

$$(K_1 \wedge K_2) \vee (K_1 \wedge K_3) = \mathbf{0}$$

Moreover, suppose that

$$A_1 \cap A_2 = \emptyset$$

and

$$A_1 \cap A_3 = \emptyset$$

for example, $A_1 = \{a, b\}$, $A_2 = \{c, d\}$, and $A_3 = \{c, e\}$. As $A_1 \cap A_2 = \emptyset$ and $A_1 \cap A_3 = \emptyset$, we have that

$$\text{tr}(A_1 \cap A_2) = T_*$$

and

$$\text{tr}(A_1 \cap A_3) = T_*$$

and therefore

$$K_1 \wedge K_2 = \mathbf{0}$$

and

$$K_1 \wedge K_3 = \mathbf{0}$$

by the definitions of minimum kind and maximum kind. It follows that

$$(K_1 \wedge K_2) \vee (K_1 \wedge K_3) = \mathbf{0}$$

since

$$\mathbf{0} \vee \mathbf{0} = \mathbf{0}$$

On the other hand, suppose

$$T_2 \cap T_3 = T_5$$

It follows that

$$\text{in}(T_2 \cap T_3) = \text{in}(T_5)$$

and thus

$$K_1 = K_1 \wedge (K_2 \vee K_3)$$

Therefore,

$$K_1 \wedge (K_2 \vee K_3) \neq (K_1 \wedge K_2) \vee (K_1 \wedge K_3)$$

One can also check that

$$(K_1 \wedge K_2) \vee (K_1 \wedge K_3) \leq K_1 \leq K_1 \wedge (K_2 \vee K_3)$$

3.3.3 Probability of Kinds

When using index terms extracted from texts, retrieval can seamlessly utilize intersection or union of the posting lists, thus implementing classical retrieval. This set-based approach to retrieval fits quite well with textual documents since the index terms are easily recognized and extracted from documents and an index term corresponds to a set of document identifiers stored in a posting list after indexing a document collection. The main assumption underlying a set-based approach to indexing, retrieval, and relevance detection is that an index term has a semantics, and its occurrence in a document is meaningful to the end users. When authors are writing their own documents, say, a_1, a_2, a_3, a_4 using, say, four index terms, namely, t_1, t_2, t_3, t_4, they assume that aboutness of documents to index terms can be expressed through the classical logical operators. Suppose, for example, that a_1, a_2, a_3, a_4 are about $\{t_1, t_2\}$, $\{t_2, t_3\}$, $\{t_2, t_3, t_4\}$, $\{t_3, t_4\}$, respectively. According to the set-based approach to IR, the posting lists $t_1 = \{a_1\}$, $t_2 = \{a_1, a_2, a_3\}$, $t_3 = \{a_2, a_3, a_4\}$, and $t_4 = \{a_3, a_4\}$ are obtained. However, the end user can utilize the operators for expressing new concepts not explicitly thought of by the document authors; for example, $t_2 \cap t_3$.

The use of kinds in IR can be explained by the need of providing the end user with a means to access information which is alternative to the traditional keyword-based means. Using kinds would help the user to express his information need using the intension or the extension of a concept.

As the retrieval of information requires ranking, in IR, kinds should be equipped with a probability distribution for ranking them against a representation of the user's information needs. Therefore, a probability function for kinds P should map a kind to the real interval $[0, 1]$. As P is a probability function, it should meet the following properties:

- P is a function of both A and T in order to exploit all the information provided by K.
- $P(K_1) \leq P(K_2)$ when $K_1 \leq K_2$ since K_1 is "smaller" than or is contained in K_2; this property is similar to the property of the probability functions applied to sets, according to which the probability of the event represented by one set is smaller than the probability of the event represented by the supersets.
- As a consequence, $P(0) \leq P(K) \leq P(1)$, since $0 \leq K \leq 1$.

- $P(0) = 0$ and $P(1) = 1$.
- It follows that when an arbitrarily large number of kinds are met, the probability of this meeting tends to 1, that is,

$$\lim_{n \to \infty} P(K_1 \vee \cdots \vee K_n) = 1$$

- Similarly, when an arbitrarily large number of kinds are joined, the probability of this joining tends to 0, that is,

$$\lim_{n \to \infty} P(K_1 \wedge \cdots \wedge K_n) = 0$$

In IR, the probabilities are usually estimated using the information provided by the posting lists of a collection index which stores information about both documents (individuals) and index terms (traits). Moreover, a probability of an index term can be a function of the number of documents indexed, while a probability of a document can be a function of the index terms stored; indeed, these statistics are exploited by the most effective weighting schemes implemented by the search engines. Something similar would be utilized to estimate the probability of kinds.

The visual description of the kinds provided by the previous figures would suggest that the areas of the rectangles could be a source to estimate the probability of kinds. If one was induced to consider the area of the rectangles drawn in Fig. 3.7 as a measure of probability, $\sum_{a \in A} \sum_{t \in T} w(a, t)$ would be used where w is the weight function given to each occurrence of a trait t in an individual a; when $w = 1$, the function is the "volume" or "area" of the kind, that is, $|A \times T| = |A||T|$ where $|\cdot|$ is the "volume" such as the cardinality of the set. However, this function does not meet the requirement that $P(1) > P(0)$. An alternative probability function is then needed.

Consider a kind $K = (A, T)$ and let $s(T) > 0$ be a function of T such that $s(T_1) \leq s(T_2)$ when $T_1 \subseteq T_2$; for example, $s(T) = |T|$. The probability of kind $K = (A, K)$ is defined as

$$P(K) = P(A)^{s(T)} \qquad 0 \leq P(A) \leq 1 \qquad 0 \leq s(T) \in \mathbb{R} \tag{3.4}$$

The observation of a kind (A, T) corresponds to the observation of the individuals of A repeated $s(T)$ times. The join and the meet of two kinds are then the conjunction and the disjunction of the outcomes of two experiments on the urn of individuals. This probability function is relevant to our purposes because:

- P is clearly a function of both A and T.
- $0 \leq P(K) \leq 1$ by the definition given by (3.4).

- When $K_1 \leq K_2$, we have that $K_1 = K_1 \wedge K_2$; it follows that

$$P(K_1) = P(K_1 \wedge K_2) = P(A_1 \cap A_2)^{s(\text{tr}(A_1 \cap A_2))}$$

since

$$P(A_1 \cap A_2) = P(A_1) \leq P(A_2) \qquad s(\text{tr}(A_1 \cap A_2)) = s(\text{tr}(A_1)) \geq s(\text{tr}(A_2))$$

we have that

$$P(K_1) = P(A_1)^{s(T_1)} \leq P(A_2)^{s(T_2)} = P(K_2)$$

- It can be noted that $P(\mathbf{1}) = P(A_*)^{s(\emptyset)} = P(A_*)^0 = 1$ and that $P(\mathbf{0}) = P(\emptyset)^{s(T_*)} = 0^{s(T_*)} = 0$.
- In the limit, meeting an arbitrarily large number of kinds yields the identity kind, thus obtaining probability 1, that is,

$$\lim_{n \to \infty} P(K_1 \vee \cdots \vee K_n) = P(\text{in}(T_1 \cap \cdots \cap T_n))^{s(T_1 \cap \cdots \cap T_n)} = 1$$

since $T_1 \cap \cdots \cap T_n$ tends to the empty set and $\text{in}(T_1 \cap \cdots \cap T_n)$ tends to A_*.
- Moreover, in the limit, joining an arbitrarily large number of kinds yields the empty kind, thus obtaining probability 0, that is,

$$\lim_{n \to \infty} P(K_1 \wedge \cdots \wedge K_n) = P(A_1 \cap \cdots \cap A_n)^{s(\text{tr}(A_1 \cap \cdots \cap A_n))} = 0$$

since $A_1 \cap \cdots \cap A_n$ tends to the empty set and $\text{tr}(A_1 \cap \cdots \cap A_n)$ tends to T_*.

As regards probabilistic IR, the difference between this probability function and the probability function adopted by the BIR model is that K denotes the individuals described by all traits in T, whereas the BIR describes the individuals as both the traits occurring and those not occurring; for example, when there are three individuals and four traits and the following table of 0/1 elements denoting nonoccurrence/occurrence of traits,

	t_1	t_2	t_3	t_4
a_1	1	1	1	0
a_2	1	1	0	1
a_3	1	1	0	1

$(\{a_1, a_2, a_3\}, \{t_1, t_2\})$ is a kind, whereas $\{a_1\}$ and $\{a_2, a_3\}$ denote two distinct events. The probability of the kind is $P(\{a_1, a_2, a_3\})^{s(\{t_1, t_2\})}$, whereas the probability of the two events are $p_1 p_2 p_3 (1 - p_4)$ and $p_1 p_2 (1 - p_3) p_4$, respectively.

Fig. 3.8 Probability and interference

K_1	K_2	$P(K_1 \vee K_2)$	Interference
(A_1, T_1)	(A_2, T_2)	$P(\text{in}(T_1 \cap T_2))^{s(T_1 \cap T_2)}$	I
(A, T)	$(A^\perp, T^\perp)$	$P(\mathbf{1})$	0
$\mathbf{0}$	(A, T)	$P(K_2)$	0
$\mathbf{1}$	(A, T)	$P(\mathbf{1})$	0

Following the postulates presented in Kolmogorov (1956), P should be such that the probability of the conjunction of two orthogonal kinds is the sum of the probabilities of the kinds. Considering K and $K^\perp$, we have that

$$P(K \vee K^\perp) = P(\mathbf{1}) + P(\mathbf{0}) = 1 + 0 = 1$$

However, this property does not hold for any pair of disjoint kinds; indeed, for $K_1 \wedge K_2 = \mathbf{0}$, we have that

$$P(K_1 \vee K_2) \neq P(K_1) + P(K_2)$$

The difference between $P(K_1 \vee K_2)$ and $P(K_1) + P(K_2)$ is due to the *superposition* of T_1 and T_2 which produces the *interference term*

$$I = P(K_1 \vee K_2) - P(K_1) - P(K_2)$$

which also exists when the kinds are disjoint and not orthogonal. Figure 3.8 summarizes the probabilities and the cases of null interference.

3.3.4   Ranking and Feedback

In this section, we introduce a method to compute kinds and two applications of kinds in IR: ranking by probability of relevance and query expansion through feedback.

3.3.4.1   Computing Kinds

The aim of a method to compute kinds is to mine all the pairs (A, T) such that each individual of A is described by all traits of T and each trait of T describes all individuals of A. The computation of the kinds of a document collection starts from the posting lists stored in the index. The problem of mining the kinds from a document collection can correspond to the problem addressed in itemset mining and illustrated by Agrawal et al. (1996), for example. If the documents of a collection (i.e., individuals) are viewed as transactions and the terms (i.e., traits) are viewed as the items, the frequent itemsets that can be mined from the collection are the subsets paper system computer (2.30961) program system computer (2.65293) program paper computer (1.81024) program paper system (1.68539) data computer system (2.40325) data program system (1.7166) programming system computer (1.77903) programming system program (1.56055) time computer system (2.46567) time program system (1.90387) time program computer (1.84145) systems computer system (2.90262) systems paper system (2.21598) systems paper computer (1.77903) systems data system (2.15356) systems time system (1.96629) systems program system (1.96629) systems program computer (1.56055) presented system computer (1.7166) presented time algorithm (1.52934) language program programming (1.68539) language paper programming (1.59176) language data programming (1.59176) information data system (1.62297) information systems system (1.65418) processing data system (1.68539) processing computer system (1.90387) processing computer data (1.62297) programs computer program (1.9975) programs system program (1.7166) programs system computer (1.59176) programs programming program (1.56055) design computer system (1.77903) design systems system (1.65418) languages language programming (2.30961) operating systems system (1.7166) output system input (1.52934) sharing system time (1.87266)

Fig. 3.9 Itemsets (with support) mined from the CACM test collection

of terms which frequently occur in the documents. The literature of itemset mining reports the two mostly used algorithms, i.e., Apriori by Agrawal et al. (1996) and FPgrowth by Han et al. (2000); for example, when FPgrowth is performed using the CACM test collection, the itemsets of Fig. 3.9 are mined.[1]

3.3.4.2 Ranking Kinds According to the PRP

In this section, we briefly describe how to apply the PRP to kinds instead of to documents. Suppose a collection of kinds has been generated from a A_* and T_*. The collection of kinds has to be split in the subcollection of kinds returned to the user because they carry relevant information and in the subcollection of kinds that are deemed nonrelevant. The Neyman and Pearson (1933) lemma can be exploited to state that retrieving the kind K is a better decision than $K^\perp$ when

$$K = \arg_{K,K^\perp} \max \mathrm{P}_1(K) - \lambda \mathrm{P}_0(K) \quad \text{where} \quad \mathrm{P}_0(K) < \alpha$$

provided that α is the maximum probability of false alarm. The split of the collection of kinds is optimal when the distance between expected recall and false alarm is maximum. The optimal split is also called "region of acceptance" using the wording of the Neyman-Pearson lemma. Following the PRP, which is derived from the lemma, the expected recall is maximum when the kinds are listed in order of $\mathrm{P}_1(K)$ and a cutoff α is applied when the given probability of false alarm is reached.

---

[1]Minimum support was 1.5 % (48 documents) and the minimum itemset size was 3.

Consider the difference between BIR and kinds mentioned in this section. According to the PRP and the Neyman-Pearson lemma, the optimal ranking criterion operates upon sets in the case of the BIR model and upon kinds in the case of this section. Therefore, the ranking criterion is always the same though the objects ranked might change from sets to kinds. The crucial difference is that sets commute when intersection, union, and complement are applied to them, whereas kinds do not when join, meet, and complement are applied to them; this is the difference that in principle helps find splits of the collection of kinds more effective than the splits of the collection of documents.

When α varies from 0 to 1 at given intervals, and a collection of kinds has to be ranked, the criterion of the Neyman-Pearson lemma arranges the kinds in order of expected recall; the kinds $K_1, \ldots, K_r$ selected up to rank $r = r(\alpha)$ give the maximum expected recall $P_1(K_1 \vee \cdots \vee K_r)$ provided that $P_0(K_1 \vee \cdots \vee K_r) < \alpha$. Note that the K_is do not necessarily commute as the events of classical IR do.

3.3.4.3 Using Kinds in Query Expansion

Consider two states of a kind, i.e., relevance and nonrelevance. Let $P_i(K)$ be the probability of K in state i where $i = 0, 1$ means nonrelevance and relevance, respectively. In particular, $P_i(K) = P_i(A)^{s_i(T)}$. The optimal region of acceptance in the case of mixed density matrices can be rewritten as the kind $K = (A, T)$, whereas the optimal region of acceptance in the case of pure density matrices can be rewritten as a new kind H. The problem is to compute H.

An approach to the problem of computing H is to take inspiration from the query expansion techniques used in IR. Suppose K has been computed by an IR system according to the PRP. From K, a series of "stretches" is performed as exemplified in Fig. 3.10; A is first withdrawn to obtain A', which is actually a subset of A; the

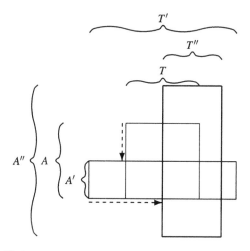

Fig. 3.10 Stretching kinds

Require: K
1: converged $\leftarrow$ FALSE
2: **while** not converged **do**
3:     $A' \leftarrow \text{distill}(A)$
4:     $T' \leftarrow \text{tr}(A')$
5:     $q'_0 \leftarrow P_0(A')^{s_0(T')}$
6:     $q'_1 \leftarrow P_1(A')^{s_1(T')}$
7:     $T'' \leftarrow \text{distill}(T')$
8:     $A'' \leftarrow \text{in}(T'')$
9:     $q''_0 \leftarrow P_0(A'')^{s_0(T'')}$
10:    $q''_1 \leftarrow P_1(A'')^{s_1(T'')}$
11:    **if** the criterion of convergence holds **then**
12:       converged $\leftarrow$ TRUE
13:    **else**
14:     $A \leftarrow A''$
15:     $T \leftarrow T''$
16:    **end if**
17: **end while**
18: $H \leftarrow (A'', T'')$

Fig. 3.11 Algorithm for stretching kinds

resulting T' is a stretch of T; T' is then withdrawn to T'', thus stretching A' to obtain A''; at each step, the kinds evolve.

Stretching kinds is formalized by the algorithm of Fig. 3.11. A subset A' of individuals is distilled from A (step 3). Document distillation can be implemented by any feedback technique used in IR; for instance, pseudo-relevance feedback techniques have been designed for distilling the candidate documents which store index terms useful for expanding the user's original query. In the case of the kinds, the index terms are given by $T' = \text{tr}(A')$ (step 4). The probabilities of correct detection and of false alarms of the new kind (A', T') are then computed at steps 5–6. Similarly to document distillation, term distillation can be implemented by any term selection used when indexing documents (step 7); for example, document parsing, structure and markup processing, link analysis, and information extraction techniques have been designed for distilling the candidate index terms useful for retrieving documents. In the case of the kinds, the documents are given by $A'' = \text{tr}(T'')$ (step 8). The probabilities of correct detection and of false alarms of the new kind (A'', T'') are then computed at steps 9–10 and compared with the probabilities of (A', T') for testing convergence; in the case of no convergence, steps 3–10 are iterated after replacing K with (A'', T'') at steps 14–15.

3.4 Concept Combination

The mathematical formalism of QM was examined and reported in the literature relevant to IR to search for the way in which words and their associations and their combinations as terms can describe meaning and what this implies for IR at the user's cognition level. The emphasis on the user's cognition level given in the literature about concept combination has been crucial since at this level, one might imagine that some quantum-like phenomena can take place, these phenomena being hardly encountered in contexts where the observables follow the classical logic and statistical behavior such as index term occurrence or other realistic properties of documents and queries.

3.4.1   Word Association

Concept combination may be acquired through free word association, and the nonrandom nature of these associations ensures that words within a vocabulary can be interconnected through special graphs which often exhibit "small worlds" within large sparse networks. The presence of "small worlds" that are small subsets of words intensely interconnected is not new in IR; it was on the contrary extensively investigated by Salton (1968), for example.

What has instead been investigated using the quantum mechanical framework is how words are represented in the user's memory as part of a network of related words since this network is crucial for allowing the user to recall a word given another word. A well-known example is given by free association. In free association, a user is presented some words, and she selects the most immediate word that comes to mind. The size of the set of associates (i.e., the world) determines the chance that a word is recalled by the user who memorized the words. Some experiments showed that the larger the set of associates, the lower the probability that an associate of the set is recalled through free association.

An interesting result stemming from these experiments was the positive correlation and the negative correlation between recall and, respectively, network density and the number of associates that occurred, although the user's attention was never drawn to the associates at any time. These experiments also showed that the denser the network linking the associated words, the higher the probability that an associate of the set is recalled through free association, the latter probability being known as recall. In other words, the user recalled the associates even though she was never stimulated to recall them after she viewed them earlier during training. It seems that the user was hearing a sort of resonance of the associates to a word when she was retrieving this word from memory.

In the literature relevant to this subject, it has been suggested that some phenomena resembling entanglement, superposition, and interference might explain the effects of word association and combination to the user's cognition; these phenomena were named "spooky activation at distance," "guppy effect," "overextension," and "underextension"; for example, the resonance of words interconnected by "small world" resembles the measurement effects on entangled pairs of photons introduced in Sect. 2.6, according to which when a measurement is performed upon a particle, other measurements performed upon that same particle at a different and very distant location are affected. The explanation provided by QM of these phenomena was implemented in terms of statistical models estimated from sets of experimental data collected through user studies.

The literature on the word combination and association based on the quantum mechanical properties of these statistical models reports that no classical statistical model can model these experimental data and that a statistical model that originated from QM is necessary. In those papers, it was attempted to link the presence of these properties to the process of concept formation, i.e., to the user's cognition level, thus leading to hypothesize that this cognition level is governed by laws similar to the laws governing atomic particles in many aspects.

According to the hypothesis that cognition level is governed by laws similar to the laws of QM, it was possible to give a formal description of other phenomena such as the disjunction effect and the conjunction fallacy in decision theory, violations of the sure-thing principle,[2] and the Allais and Ellsberg paradoxes in economics,[3] which, although not directly relevant to IR, may give some hints about possible, less traditional views of the information access processes.

The key notion of the idea of using the quantum mechanical framework in explaining the genesis and the impact of word combination and association was the notion of context, which is constantly occurring in IR. The words to be combined and associated are continuously changed by context. The change or evolution of concepts can be paralleled to the change or evolution of the state of an atomic particle subjected to a measurement. The connection to the physics of atomic particles is also explained by the fact that this physics is essentially under the influence of the physical context which continuously affects the measurement instruments and then the measures observed in an atomic particle. In physics, the presence of the quantum mechanical nature of a phenomenon can be proved by looking at experimental data.

The main assumption made by the researchers who investigated the quantum mechanical nature of the user's cognition level was that it is irrelevant to the validity of this investigation whether the data are the result of experiments in physics or in any other domain of science, that is, what happens to the user's cognition level can straightforwardly be put in correspondence with what happens in the physical atomic world.

---

[2] A posteriori, an agent would have made the same choice regardless of the events which could have occurred in the meantime.

[3] An agent prefers taking on risks where he knows the associated probabilities and costs rather than risks where it is impossible to estimate the associated probabilities or costs.

3.4.2 State, Context and Property

Aerts and Czachor (2004) and Aerts and Gabora (2004a) introduced the State, Context and Property (SCOP) formalism to define concepts using the following definitions:

- An exemplar was any entity with distinct and independent existence serving as a typical example of a class or a concept; for example, "apple" and "guppy" were exemplars.
- A concept was a class of exemplars gathered together because they have some predefined characteristic in common and differentiated from others by other characteristics; for example, *fruit* was a class and "apple" was an exemplar of this class; *pet* was a class and "guppy" was an exemplar of this class.
- A context was a sentence written in natural language providing a meaning to an exemplar depending on the concept of the exemplar and on the context thereof; for example, "the *pet* is chewing a bone" is a context and "the *fruit* gets squeezed for a fresh drink of juice" is another context. The special ground context "the *pet* is a *pet*" was also given.
- A property is a proposition which is either true or false depending both on the context and on the concept (not on the exemplar); for example, the property "it is feathered" is true when it is applied to *pet* in the context "the *pet* is being taught to talk."

The quantum formalism was utilized to describe the notions above listed. In particular:

- A concept was corresponded to a state and therefore was described as a density matrix. Concepts can be combined; for example, *pet* and *fish* can be combined to obtain *pet fish* which is in a different state from the states of the constituent concepts. These combinations are viewed as entangled photons since the outcomes of the measurement of an observable in each constituent concepts are correlated, thus giving rise to the "guppy effect."
- A context was corresponded to an observable and therefore was described as a set of projectors and specifically yes/no projectors. These observables were named "context" to stress the dynamic nature of the state of a concept. A concept is an entity that can be in different states because a context (i.e., an observable) causes a change of state of the concept; for example, the concept *pet* is in a state in the context "the *pet* is chewing a bone" and it is in another state in the context "did you see the type of *pet* he has? This explains that he is a weird person."[4]

---

[4]These examples are in Aerts and Czachor (2004) and Aerts and Gabora (2004a)'s papers.

3.4.3 Using Superposition to Define Concepts

Experiments that suggested the use of the statistical models commonly used in QM were carried out in the 1980s and resulted in the so-called guppy effect in concept combination. In these experiments, the concepts (or classes) *pet*, *fish*, and *pet fish* were defined, and some words were observed. It was found that words such as "guppy" were a very typical example of *pet fish*, but it was neither an example of *pet* nor of *fish*.

In IR, the difference between terms like *pet fish* and terms built by conjunction like *pet* AND *fish*, the latter referring to the intersection between the classes *pet* and *fish*, was often found. A guppy is not a typical pet, nor is a guppy a typical fish, but a guppy is a typical pet fish. As another example, when combining concepts, e.g., *stone* and *lion* to produce *stone lion*, however, the only aspect of the object that is lionlike is its shape. One cannot conclude that a stone lion is alive; this effect is called non-monotonicity of the combined concept.

Besides the "guppy effect," other experiments found two other phenomena, i.e., overextension and underextension. The following is an example of overextension. The subjects who participated in some experiments were asked to decide the membership of some words to the classes *bird* and *pet*. The subjects were allowed and asked to use measures of fuzzy membership ranging between 0 and 1. The experiments reported that "cuckoo" was a member of *bird* with weight 1 and was a member of *pet* with weight 0.575. The surprising result was that the subjects weighted 0.875 the membership of "cuckoo" to the intersection *pet* AND *bird*, that is, it was more probable that "cuckoo" is a member of the intersection than the individual classes.

However, if the strength of the membership of a member to a class is measured using probability, the Kolmogorov axioms will dictate that

$$P(\text{pet} \wedge \text{bird}) \leq \min\{P(\text{pet}), P(\text{bird})\} \tag{3.5}$$

As for underextension, the classes *home furnishings* and *furniture* and the disjunction *home furnishings* OR *furniture* were proposed to the subjects who were asked to place the word "ashtray." Subjects rated the membership weight of "ashtray" for the classes *home furnishings* and *furniture* as 0.7 and 0.3, respectively, but the membership weight for the disjunction was rated 0.25. The probability that "ashtray" was a member of a disjunction was less than the minimum between the probabilities that it was a member of a class, that is, there was a violation of the inequality that

$$P(\text{home furnishings} \vee \text{furniture}) \geq \max\{P(\text{home furnishings}), P(\text{furniture})\} \tag{3.6}$$

The phenomena that are known as "guppy effect," overextension, and under-extension can be described using the algebraic quantum mathematical framework. Consider the classes *pet* and *bird* which are both described as the states $|\text{pet}\rangle$ and $|\text{bird}\rangle$ of a vector space. Their disjunction can be represented as the superposition of the states which represent the classes and can be written as

$$\frac{1}{\sqrt{2}}|\text{pet}\rangle + \frac{1}{\sqrt{2}}|\text{bird}\rangle$$

Consider the observable such that a word belongs to a class. Following the axioms explained in Sect. 2.4, the probability that a word belongs to the class represented by $|\text{pet}\rangle$ is

$$\text{tr}(|\text{pet}\rangle\langle\text{pet}|w\rangle\langle w|) = \langle\text{pet}|w\rangle\langle w|\text{pet}\rangle = |\langle w|\text{pet}\rangle|^2$$

where $|w\rangle\langle w|$ is the projector corresponding to the event that a word belongs to the class. It follows that

$$\frac{1}{2}|\langle w|\text{pet}\rangle|^2 + \frac{1}{2}|\langle w|\text{bird}\rangle|^2 + I$$

is the probability that a word belongs to the superposition of the classes where I is the interference term (Sect. 2.5), which ranges between -2 and $+2$. The interference term allows us to violate the inequalities (3.5) and (3.6) imposed on probabilities.

As the experimental data reported in the literature violates the inequalities (see the readings suggested in Sect. 3.12), it is argued that these data cannot result from a classical space, but can only result from a quantum mechanical space. This argument suggests that qubits may be a useful device for representing the user's cognition level. Therefore, suppose that the user's cognition level can be represented using qubits where one qubit represents the state of a word in the cognition level of the user; for example, these words may be query words to be submitted to an IR system. The ground states of a qubit can in this case correspond to the event that a word has been chosen by the user or that it has not been chosen. Thus, if we represent the words using the standard superpositions

$$|v\rangle = a_0|0\rangle + a_1|1\rangle \qquad |w\rangle = b_0|0\rangle + b_1|1\rangle$$

it is possible to denote the state of the combined concept using the tensor product

$$|v\rangle \otimes |w\rangle = a_0 b_0|00\rangle + a_1 b_0|10\rangle + a_0 b_1|01\rangle + a_1 b_1|11\rangle$$

where $|a_0 b_0|^2 + |a_1 b_0|^2 + |a_0 b_1|^2 + |a_1 b_1|^2 = 1$ or using entangled states which cannot be decomposed as explained in the next section.

3.4.4   Using Entanglement to Define Concepts

Concept combination can also be modeled by entanglement. A concept is a combination of two entangled words (or concepts) when it cannot be separated, that is, the meaning of the concept cannot be reconstructed from the meaning of the individual component words. As an example of an entangled state, we might consider the state

$$\frac{1}{\sqrt{2}}|00\rangle + \frac{1}{\sqrt{2}}|11\rangle$$

where the words are either both chosen or both not chosen by the user. If the state of a concept is entangled, it cannot be separated and one cannot express it saying that, for example, one word has been chosen and the other has not.

In QM, the criteria used to test entanglement are given by the Bell inequalities. A possible way to proceed is to define four observables. Each observable is binary and thus gives two mutually exclusive outcomes; for example, an observable can tell whether a word has been chosen or not. Each pair of observables is applied to an instance (e.g., "guppy") of a combined concept (e.g., *pet fish*): +1 is recorded when the observables give concordant outcomes (i.e., either two ones or two zeros) or −1 is recorded when the observables give discordant outcomes. When each pair of observables i, j is applied several times, it is possible to estimate the following expectation:

$$E(i,j) = +1\mathrm{P}(i,j \text{ concordant}) + (-1)\mathrm{P}(i,j \text{ discordant})$$

If the concept under observation is separable, the following inequality holds:

$$|E(1,2) - E(1,4)| + |E(2,3) + E(2,4)| \leq 2 \qquad (3.7)$$

The violation of (3.7) is a signal of entanglement.

The possibility of expressing concept combination using quantum entanglement prompts the question whether entanglement is applicable to the cognition level of a user of an IR system. If this was possible, that is, if a large series of experiments confirmed the existence of entanglement during the interaction between users and IR systems, it would be possible to design systems which can take into account the fact that there are combined concepts that cannot be easily expressed by saying that a word can be chosen by the user. Testing for entanglement at the user's cognition level requires the use of some testable criteria which can be summarized as follows: if the concept is separable (i.e., not entangled), then it will satisfy the criteria, that is, if the criteria are not satisfied, the concept is entangled. However, it seems very hard to design experiments that put the user's cognition level in an entangled state.

3.4.5 Adding Superposition to Language Modeling

Suppose a concept can be implemented by a document or a collection of documents. In particular, suppose a document B implements a concept. The distribution of the probability of occurrence of a word s is assigned to this concept in the sense that the word s occurs with probability provided by the distribution estimated by the data extracted from the document. The occurrence of a word is an observable, and the distribution of probability can be described by a density matrix ρ_B, that is,

$$P(s \text{ occurs in } B) = \text{tr}(\rho_B |s\rangle \langle s|) = \langle s|\rho_B|s\rangle$$

If the probability distribution is in particular given by the state vector $|\psi\rangle$, we have that

$$P(s \text{ occurs in } B) = |\langle s|\psi_B\rangle|^2$$

For example, consider document FBIS4-7688 of the TIPSTER test collection and the query word "organized" of topic 301; the query built from the title and the description of this topic includes other words and the total query word frequency in the document, that is, the sum of the single query word frequencies in the document is 31. As "organized" occurs five times, the probability of occurrence of this word can be estimated to be $5/31$. We may then write

$$|\psi_B\rangle = \begin{pmatrix} \sqrt{\frac{5}{31}} \\ \sqrt{\frac{26}{31}} \end{pmatrix} \qquad \rho_B = \begin{pmatrix} \frac{5}{31} & \frac{\sqrt{130}}{31} \\ \frac{\sqrt{130}}{31} & \frac{26}{31} \end{pmatrix}$$

The document used to implement a concept usually belongs to a collection or in general to some groups of documents, each group being created according to some predefined criteria (e.g., the same topic). A question frequently asked in IR is how probable is the generation of a word by a document. This is indeed the question asked within the LM approach described in Sect. 1.4.2; the answer to this question is the measure used to rank a document returned to the end user who submitted a query containing the word. The idea that underlies the LM approach is that the word may occur either in the document B or in the group L of documents including B. This idea can be seen from the formula (1.3) used to calculate the probability of occurrence of s using the LM approach and derived from the idea that s is generated either by B (with probability $1 - \lambda$) or by L (with probability λ). Indeed, we have that

$$P(w \text{ occurs in } B) = \frac{f(s_{(j)}, B)}{\sum_{j=1}^{n} f(s_{(j)}, B)}$$

$$P(w \text{ occurs in } L) = \frac{f(s_{(j)}, L)}{\sum_{j=1}^{n} f(s_{(j)}, L)}$$

$$P(B) = 1 - \lambda$$
$$P(L) = \lambda$$

thus obtaining that

$$\hat{p}_B(s_{(j)}) = P(w \text{ occurs})$$
$$= P(w \text{ occurs in } B \text{ or } w \text{ occurs in } L)$$
$$= P(w \text{ occurs in } B)P(B) + P(w \text{ occurs in } L)P(L)$$

which is equal to expression (1.3). This formulation means that B and L are two mutually exclusive urns (i.e., language models) in the sense that s can be observed from either B or L. Using the quantum mechanical framework, (1.3) can be written as a mixture as follows:

$$(1 - \lambda)\text{tr}(\rho_B |s\rangle\langle s|) + \lambda \text{tr}(\rho_L |s\rangle\langle s|)$$

However, using this framework, the probability of occurrence of s can also be calculated by a superposition between B and L. Conceptually speaking, such a superposition is different from the mixture used within the LM approach. While the mixture within the LM approach indicates that the occurrence of s in B is somehow "separated" from the occurrence of s in L, a superposition would indicate that B may be placed above L or vice versa and that the occurrence of s in B can also be viewed as an occurrence in L or vice versa, thus suggesting a sort of interference between the two occurrences.

Using the quantum mechanical framework, the distribution of probability of the occurrence of s can be expressed using a density matrix ρ corresponding to a state vector $|\psi\rangle$ which is a superposition of the state vectors $|\psi_B\rangle$ and $|\psi_L\rangle$, the former being the representation of the distribution of probability of the occurrence of s in B and the latter being the representation of the distribution of probability of the occurrence of s in L. This superposition can be written as follows:

$$|\psi\rangle = \sqrt{1 - \lambda}|\psi_B\rangle + \sqrt{\lambda}|\psi_L\rangle$$

Note that the probability of B is again the same calculated using the mixture, that is,

$$1 - \lambda = P(B) = \text{tr}(\rho|\psi_B\rangle\langle\psi_B|) = |\langle\psi|\psi_B\rangle|^2$$

but the probability of the occurrence of s differs from the probability calculated using the mixture because

$$P(s \text{ occurs}) = \text{tr}(\rho|s\rangle\langle s|)$$
$$= \langle s|\rho|s\rangle$$
$$= \langle s|\psi\rangle\langle\psi|s\rangle$$

$$= \langle s|(\sqrt{1-\lambda}|\psi_B\rangle + \sqrt{\lambda}|\psi_L\rangle)(\sqrt{1-\lambda}\langle\psi_B| + \sqrt{\lambda}\langle\psi_L|)|s\rangle$$
$$= (\sqrt{1-\lambda}\langle s|\psi_B\rangle + \sqrt{\lambda}\langle s|\psi_L\rangle)(\sqrt{1-\lambda}\langle\psi_B|s\rangle + \sqrt{\lambda}\langle\psi_L|s\rangle)$$
$$= (1-\lambda)|\langle s|\psi_B\rangle|^2 + \lambda|\langle\psi_L|s\rangle|^2 + 2I$$
$$= (1-\lambda)P(s \text{ occurs in } B) + \lambda P(s \text{ occurs in } L) + 2I$$

where

$$I = \sqrt{\lambda}\sqrt{1-\lambda}\langle s|\psi_L\rangle\langle\psi_B|s\rangle \cos\theta$$

is the interference term and $\cos\theta$ (i.e., θ) is an additional parameter to be estimated. The estimation of $\cos\theta$ raises issues similar to those raised by the estimation of λ and of the (a, b)s in the event that smoothing is used within the LM approach as described by (1.4). Similar to the strategies used to estimate λ and of the (a, b)s, a strategy to estimate $\cos\theta$ can be a search of a number of the possible values of $\cos\theta$ in the range $[-1, +1]$ and then the selection of the value which maximizes a certain retrieval effectiveness measure. However, such a search should be performed for each word, thus making this strategy little practical. An alternative strategy would consider the following calculation:

$$(1-\lambda)P(Q|B) + \lambda P(Q|L) + 2I$$

where

$$P(Q|B) = \text{tr}(\rho_B|Q\rangle\langle Q|)$$

is the probability of Q using the QLM estimated by B and

$$P(Q|L) = \text{tr}(\rho_L|Q\rangle\langle Q|)$$

is the probability of Q using the QLM estimated by L. When $\cos\theta$ is different from zero, the probability of Q calculated by the superposition between B and L is different from the probability of Q calculated by the (classical) mixture of B and L determined by the QLM.

According to the QLM, the event that Q is generated by a mixture of B and L is a signal that B is relevant to the information need represented by Q. When superposition is utilized as described in this section, the estimation of the interference term is needed. The estimation of $\cos\theta$ should consider the meaning of the interference term in the context of the LM approach to IR. If we borrow the metaphor of the double-slit experiments illustrated in Sect. 2.5, it is possible to describe the document B and the group of documents L as slits, and the interference between slits can be used to describe the interference between B and L. When the interference is positive ($\cos\theta > 0$), the probability of the arrival of the particle at the shield is higher than the probability provided by the mixture of the probabilities of passage through each slit; it is like an "amplification" caused by the fact that both slits are open impacts on the event that the particles arrive at the shield. Therefore, $\cos\theta$ should be positive when some additional evidence about the relevance of B will be available other than the evidence utilized in the estimation of the probabilities. In a similar way, when the interference is negative ($\cos\theta < 0$), the probability of Q is smaller than the probability provided by the mixture of the probabilities of Q in B and of Q in L; $\cos\theta$ should be negative when some additional evidence about the irrelevance of B will be available. Examples of additional evidence that may make the document preferable to the documents of the group are click-through data (the more the document is clicked, the higher the probability of relevance), data quality (the more the document is written and orderly organized, the more the document can be appreciated by the user), and user's task (the more the document meets the user's task, the more the document will be deemed useful).

3.5   Word Ambiguity

Ambiguity is a key problem in IR both when a word can have more than one meaning (polysemy) and when a word can convey meaning better than another word in a certain context (synonymy). In particular, ambiguity affects textual document retrieval systems (e.g., commercial search engines and speech retrieval systems) and in general the systems dealing with text representations. In this section, we explain how the problem can be described by some structures of the theory of QM.

3.5.1   Ambiguity and Superposition

The connection between word ambiguity and the superposition of atomic particle states is based on the linear combination of basis vectors. Superposition can indeed be represented by a linear combination of the vectors representing the pure (or ground) states. The vector that results from a linear combination represents a state of the particle which is a superposition of pure states, i.e., that the particle is not in any state out of the pure states since it is on the contrary in a state which cannot be included in the basis of pure states. It is only when the particle can be measured that its state collapses to one of the pure states, and the state vector collapses to the corresponding basis vector. The probability that the particle state collapses to one of the pure states is the squared modulus of the coefficient used for linearly combining the resulting basis vectors with the other basis vectors.

Word ambiguity is represented in a way similar to the way superposition is represented. The ambiguity of a word can indeed be represented by a linear combination of the vectors representing the distinct known meanings of the word. The vector that results from a linear combination represents an ambiguous word which is a superposition of known meanings, thus meaning that the word does not possess any meaning out of the known ones since it is on the contrary in a state which cannot be included in the basis of known meanings. It is only when the word can be observed within a certain context that its ambiguous meaning collapses to one of the known meanings and the word vector collapses to the basis vector corresponding to the known meaning.

The probability that the ambiguous word collapses to one of the known meanings is the squared modulus of the coefficient used for linearly combining the resulting basis vectors with the other basis vectors; for example, suppose the ambiguous word "java" is represented by the vector $|java\rangle$. The ambiguity of "java" can be represented by

$$|java\rangle = a|api\rangle + b|island\rangle + c|coffee\rangle \tag{3.8}$$

where

$$|a|^2 = P(java \text{ has the meaning api})$$

$$|b|^2 = P(java \text{ has the meaning island})$$

$$|c|^2 = P(java \text{ has the meaning coffee})$$

and

$$|a|^2 + |b|^2 + |c|^2 = 1$$

The three vectors $|api\rangle$, $|island\rangle$, and $|coffee\rangle$ may refer, for example, to three documents, respectively, about the Java Application Programming Language (API), the Java island bay, and the Java coffee. These three document vectors must be orthogonal to make probability computation possible and to describe the fact that when measured, the meaning gives only one value chosen from API, island, and coffee. The amplitudes a, b, c can be estimated by the number of times the ambiguous word is used with each of its possible meanings in a reference corpus of documents. It should be noted however that these frequencies provide real numbers, which are indeed the modula of the amplitudes, whereas the amplitudes are actually complex numbers; however, the complex field to which a, b, c belong has not been investigated in the literature of this subject. Figure 3.12 provides a

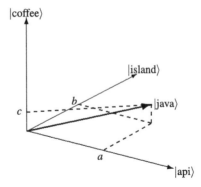

Fig. 3.12 Ambiguity and superposition

pictorial description of the correspondence between ambiguity and superposition; |java⟩ is represented by a vector spanned by the three basis vectors according to a, b, c; the closer the vector |java⟩ is to a basis vector, the higher the corresponding coefficient. As the coefficients a, b, c vary, the vector |java⟩ spans a unit-radius sphere in the three-dimensional space. It should be noted that the vector |java⟩ always has unitary length since the coordinates are amplitudes. As the three meaning vectors are mutually orthogonal, that is,

$$\langle api|island \rangle = 0$$

$$\langle api|coffee \rangle = 0$$

and

$$\langle coffee|island \rangle = 0$$

and are unitary, the squared amplitudes are probabilities which sum to one.

Word disambiguation might therefore be modeled using the orthogonality between the basis vector. To obtain the meaning given by API, |java⟩ can be projected to the subspace spanned by |api⟩, and the projection is actually this basis vector. To obtain the representation of the meaning of java without the meaning given by API, |java⟩ can be projected to the subspace spanned by |island⟩ and |coffee⟩ implemented by the projector |island⟩⟨island| + |coffee⟩⟨coffee|, thus obtaining

$$b|island \rangle + c|coffee \rangle$$

which belongs to the bidimensional subspace spanned by |island⟩ and |coffee⟩ and orthogonal to the subspace spanned by |api⟩.

3.5.2 *Vector Negation and Negative Relevance Feedback*

An alternate approach to disambiguation is through vector negation. This technique was proposed by Widdows and Peters (2003) and later explained in the book of Widdows (2004) together with other concepts of the quantum mechanical framework. In the following, vector negation is first introduced in the context of negative RF, and then the commutativity of vector negation will be discussed.

In Sect. 1.3, RF was introduced for the VSM, and two types of feedback were introduced, positive feedback and negative feedback. The effectiveness of RF lies on the hypothesis that the relevant document vectors tend to be closer to each other than to the nonrelevant document vectors; this hypothesis is known as cluster hypothesis.

Negative RF is problematic because the cluster hypothesis does not hold for nonrelevant documents, that is, it cannot be argued that the nonrelevant document vectors tend to be closer to each other than to the relevant document vectors. If it did, it would be possible to split the space of the document vectors into two clusters, a cluster of relevant documents and a cluster of nonrelevant documents.

Negative RF might be addressed in a more principled way to attempt to overcome the problems due to the invalidity of the cluster hypothesis. Such an approach is based on vector negation. Vector negation models logical negation using the orthogonality between vectors. Consider the vector $|\text{array}\rangle$, $|\text{bank}\rangle$, and $|\text{credit}\rangle$, respectively, for the three terms "array", "bank," and "credit":

$$|\text{array}\rangle = \begin{pmatrix} 1 \\ 0 \\ 0 \end{pmatrix} \qquad |\text{bank}\rangle = \begin{pmatrix} 0 \\ 1 \\ 0 \end{pmatrix} \qquad |\text{credit}\rangle = \begin{pmatrix} 0 \\ 0 \\ 1 \end{pmatrix}$$

The term "bank" is ambiguous since it has many different meanings in a natural language. One meaning of "bank" might be an array of similar things, especially electrical or electronic devices, grouped together in rows. Another meaning might be that of a financial establishment that uses money deposited by customers for credit, investment, and currency exchange. Suppose there are two documents. Document d_1 is about bank as array and document d_2 is about bank as credit. The document about bank as array can be represented by the following ket:

$$|d_1\rangle = |\text{array}\rangle + 2|\text{bank}\rangle = \begin{pmatrix} 1 \\ 2 \\ 0 \end{pmatrix}$$

whereas the document about bank as credit can be represented by the following ket:

$$|d_2\rangle = |\text{bank}\rangle + |\text{credit}\rangle = \begin{pmatrix} 0 \\ 1 \\ 1 \end{pmatrix}$$

Note that d_1 is about bank but not about credit since

$$\langle d_1 | \text{credit} \rangle = 0$$

and d_2 is about bank but not about array since

$$\langle d_2 | \text{array} \rangle = 0$$

Suppose the query "bank as an array" is submitted by an end user to an IR system. Such a query can be represented by the following vector:

$$|q\rangle = |\text{array}\rangle + |\text{bank}\rangle = \begin{pmatrix} 1 \\ 1 \\ 0 \end{pmatrix}$$

The system will rank the documents by the inner products between the query vector and the document vectors calculated as follows:

$$\langle d_1 | q \rangle = 3 \qquad \langle d_2 | q \rangle = 1$$

Suppose instead the user wants to submit the query "bank" but he is actually thinking about bank as array; this is indeed an ambiguous query. The vectorial representation of the query would then be

$$|q'\rangle = |\text{array}\rangle = \begin{pmatrix} 0 \\ 1 \\ 0 \end{pmatrix}$$

and the system will compute the following inner products:

$$\langle d_1 | q' \rangle = 1 \qquad \langle d_2 | q' \rangle = 1$$

The document d_2 is about credit and should be removed from the list of the retrieved documents. To this end, the user should be able to express the query "bank" AND NOT "credit"; however, the system is working according to the VSM which does not provide Boolean operators.

To address this lack of operators, it is possible to leverage the RF of the VSM and to refine it to the query vector in order to remove the unwanted documents from the list of the retrieved documents. Consider the way the VSM works. When a user adds terms to describe his information need, the system based on the VSM adds term vectors to the query vector. Suppose two query terms t_1, t_2 describe the need and the user wants all and only the documents indexed by both terms. To this end, he will submit a query like t_1 AND t_2. Suppose the user no longer wants the documents

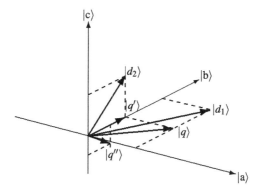

Fig. 3.13 Vector negation and negative RF

about t_2. To this end, if t_2 were no longer describing the need, then t_1 AND NOT t_2 would be the right query.

According to the VSM, the term vectors should be subtracted from the query vector. This subtraction is actually negative RF; however, the negative RF of the VSM requires that the β parameters have to be defined precisely. Although the VSM tells what to do with the vectors to implement negative feedback, it does not provide insights on how to define the parameters.

A possible approach to defining the parameters of the vectors subtracted by negative feedback is algebraic and depicted in Fig. 3.13. The new query vector should be

$$|q''\rangle = |q'\rangle - \beta|d_2\rangle \qquad (3.9)$$

with the constraint that

$$\langle d_2|q''\rangle = 0$$

since the document d_2 is nonrelevant to q''. To obtain β, (3.9) is multiplied by $\langle d_2|$ and set to zero so that

$$\langle d_2|q''\rangle = 0 \quad \text{if and only if} \quad \beta = \frac{\langle d_2|q'\rangle}{\langle d_2|d_2\rangle}$$

Indeed, it can be checked that

$$\langle d_2|q''\rangle = \langle d_2|(|q'\rangle - \beta|d_2\rangle)$$
$$= \langle d_2|q'\rangle - \beta\langle d_2|d_2\rangle$$
$$= \langle d_2|q'\rangle - \langle d_2|q'\rangle$$
$$= 0$$

and that

$$\langle d_1|q''\rangle = \langle d_1|(|q'\rangle - \beta|d_2\rangle) = \langle d_1|q'\rangle - \beta\langle d_1|d_2\rangle$$

Consider the example above, that is,

$$|q''\rangle = \begin{pmatrix} 0 \\ 1 \\ 0 \end{pmatrix} - \beta \begin{pmatrix} 0 \\ 1 \\ 1 \end{pmatrix}$$

We have that $\beta = \frac{1}{2}$; therefore,

$$|q''\rangle = \begin{pmatrix} 0 \\ \frac{1}{2} \\ -\frac{1}{2} \end{pmatrix}$$

Indeed

$$\langle d_2|q''\rangle = 0$$

as requested.

Vector negation is applied to the case that the word vectors (e.g., $|java\rangle$) are spanned by basis document vectors which provide the possible meanings of the word. Suppose java is represented by the superposed state vector (3.8) and suppose it is stripped of both the meanings represented by $|api\rangle$ and $|island\rangle$. Using vector negation, the vector of the word stripped of both the meaning given by $|api\rangle$ and the meaning given by $|island\rangle$ is obtained as follows. First, the vector depurated of the meaning of API is calculated:

$$|java\_no\_api\rangle = |java\rangle - \frac{\langle api|java\rangle}{\langle api|api\rangle}|api\rangle$$

For the sake of clarity, suppose the vectors have unitary length:

$$|java\_no\_api\rangle = |java\rangle - \langle api|java\rangle|api\rangle$$

Then, the vector depurated of the meaning of "island" is calculated from the vector depurated of the meaning of API:

$$|java\_no\_api\rangle - \langle island|java\_no\_api\rangle|island\rangle$$

If the right-hand side of the latter replaces $|java\_no\_api\rangle$ in the earlier subtraction, we have that

$$|java\rangle - \langle api|java\rangle|api\rangle - (\langle island|java\rangle - \langle api|java\rangle\langle island|api\rangle)|island\rangle$$

Suppose $|\text{island}\rangle$ is subtracted first and $|\text{api}\rangle$ is subtracted second. The resulting vector is

$$|\text{java}\rangle - \langle\text{island}|\text{java}\rangle|\text{island}\rangle - (\langle\text{api}|\text{java}\rangle - \langle\text{island}|\text{java}\rangle\langle\text{api}|\text{island}\rangle)\,|\text{api}\rangle$$

The representations of the words stripped of two meanings subsequently removed are equivalent only under some simplifying assumptions. If these assumptions between the three basis vectors cannot be made, the subtractions interfere with each other, thus meaning that two successive negations represented in vector spaces are different from two successive negations represented by sets.

The invalidity of these assumptions about the vectors that represent word meaning gives an additional degree of freedom to the design of IR systems since it allows the designer to tailor the document ranking resulting from a subtracted word (e.g., from a query) according to the order of subtraction. However, this causes an additional degree of complexity to the designer of experiments who has to consider the possible orders of subtraction when testing whether vector negation is effective and what order of subtraction is the most effective.

3.6 Semantic Spaces

The use of the QM notions in some contexts of artificial intelligence has been based on conceptual spaces and reported since Bruza and Cole (2005)'s work. A conceptual space can be defined as a space where objects are categorized; for example, a document is about a topic or a term is an instance of a concept. A conceptual space is thus a space where objects and classes are related together. A semantic space is a computational approximation of a conceptual space.

3.6.1 Hyperspace Analogue to Language

A common implementation of a semantic space is provided by an algorithm called Hyperspace Analogue to Language (HAL) which resembles many algorithms used in IR to compute similarities between words or documents. Given a set of textual documents and the vocabulary of k distinct words occurring in the documents, HAL calculates a $k \times k$ matrix where an element of the matrix is a measure of word similarity. A word similarity measure is based on proximity.

The matrix can be constructed by moving a window of a given length over the set of textual documents, document by document, by one word increment. All words within the window of the given length are considered as co-occurring with the last word or the central word in the window with a strength inversely proportional to

```
Require: D,w
  HAL(D,w)
  k is the number of distinct words
  H is a k × k matrix
  for all d ∈ D do
    for i = 1,...,length(d) do
      for j = 1,...,max{i+1,length(d) − w} do
        h_{ij} ← h_{ij} + closeness(d_i,d_j)
        h_{ji} ← h_{ij}
      end for
    end for
  end for

  return H/wk
```

Fig. 3.14 Hyperspace Analogue to Language

the distance between the words. The algorithm of HAL is reported in Fig. 3.14. D is a subset of documents, w is the window size, and $\mathbf{H}$ is the matrix computed by HAL. Suppose a closeness function is defined to compute the closeness between two words d_i, d_j of a document d; for example, $\text{closeness}(d_i, d_j) = w + 1 - (i - j)$; the algorithm returns a matrix which is symmetric because it needs to be a Hermitian matrix over the real field for implementing a density matrix; for example, when the document s is "Extraction of Roots by Repeated Subtractions for Digital Computers", the closeness function is $w + 1 - (i - j)$ and $w = 2$; we have that

$$
\mathbf{H} = \frac{1}{18}
\begin{pmatrix}
3 & 2 & 1 & 0 & 0 & 0 \\
2 & 3 & 2 & 1 & 0 & 0 \\
1 & 2 & 3 & 2 & 1 & 0 \\
0 & 1 & 2 & 3 & 2 & 1 \\
0 & 0 & 1 & 2 & 3 & 2 \\
0 & 0 & 0 & 1 & 2 & 3
\end{pmatrix}
\begin{matrix}
\text{extract} \\
\text{root} \\
\text{repeat} \\
\text{subtraction} \\
\text{digital} \\
\text{computer}
\end{matrix}
$$

The highest closenesses are between each word and itself and decrease when the distance between two words increases until it becomes null because a word is "out of window."

Another way to compute a semantic space is to calculate the word similarity matrix from a document collection after filtering the documents using a query; in this way, the filtered document subset will be rich of relevant documents. Consider, for example, the CACM test collection designed and implemented by Fox (1983). Suppose the documents are filtered by the following query:

What articles exist which deal with TSS Time Sharing System, an operating system for IBM computers?

in a way that only the documents of the collection that are indexed by at least one query word are selected. Word similarity has then been computed as the cosine between the word vectors expressed as vectors of the frequency of a word relative to the document lengths. The cosines of the word pairs are reported in the following matrix:

	articles	computers	deal	exist	ibm	operating	sharing	system	time	tss
articles	1.000	0.000	0.000	0.000	0.004	0.000	0.000	0.002	0.000	0.000
computers	0.000	1.000	0.001	0.001	0.001	0.001	0.001	0.001	0.001	0.000
deal	0.000	0.001	1.000	0.006	0.000	0.003	0.001	0.002	0.001	0.000
exist	0.000	0.001	0.006	1.000	0.000	0.003	0.000	0.001	0.000	0.000
ibm	0.004	0.001	0.000	0.000	1.000	0.001	0.001	0.003	0.000	0.000
operating	0.000	0.001	0.003	0.003	0.001	1.000	0.003	0.005	0.001	0.000
sharing	0.000	0.001	0.001	0.000	0.001	0.003	1.000	0.004	0.004	0.012
system	0.002	0.001	0.002	0.001	0.003	0.005	0.004	1.000	0.002	0.003
time	0.000	0.001	0.001	0.000	0.000	0.001	0.004	0.002	1.000	0.006
tss	0.000	0.000	0.000	0.000	0.000	0.000	0.012	0.003	0.000	0.006

This matrix is an implementation of a semantic space whereby the dimensions of the space correspond to words. Each dimension represents the "state" of the word in the context of the CACM text collection and of the query from which the semantic space was computed. As the "state" of the word is implemented by a column vector, it might be viewed as a state vector as the vector used in QM to describe the state of a particle.

Similar to the changes of a state vector when the context changes, if the collection or the query changes, the state vector of the word may also change. In this case, the state of a word in a semantic space is related to meaning, and this meaning is a dimension of context. Therefore, the changes to the collection or to the query influence the meaning of the word and therefore its context.

The dimensions that can be derived from the context given by the collection and the query can be calculated by the SVD of the word similarity matrix. The eigenvalues are as follows:

$$1.013\ 1.005\ 1.003\ 0.999\ 0.998\ 0.995\ 0.995\ 0.994\ 0.993\ 0.006$$

The eigenvectors are as follows:

articles	+0.16	+0.43	−0.49	+0.14	−0.24	+0.52	−0.35	+0.06	−0.26	+0.00
computers	+0.16	+0.06	+0.02	−0.97	+0.07	+0.07	−0.16	+0.02	−0.05	+0.00
deal	+0.35	−0.49	−0.21	−0.02	−0.24	+0.09	+0.09	−0.71	−0.07	+0.00
exist	+0.30	−0.52	−0.28	−0.02	−0.26	+0.03	+0.13	+0.68	+0.13	+0.00
ibm	+0.23	+0.44	−0.43	−0.07	−0.11	−0.57	+0.47	−0.04	−0.03	+0.00
operating	+0.46	−0.12	+0.04	+0.17	+0.52	−0.30	−0.32	+0.09	−0.52	+0.00
sharing	+0.38	+0.18	+0.48	+0.04	−0.04	+0.41	+0.58	+0.08	−0.28	+0.01
system	+0.52	+0.20	+0.03	+0.11	+0.31	+0.12	−0.13	−0.07	+0.74	+0.00
time	+0.26	+0.14	+0.46	+0.05	−0.66	−0.34	−0.39	+0.02	+0.03	+0.00
tss	+0.01	+0.00	+0.01	+0.00	+0.00	+0.01	+0.01	+0.00	−0.00	+1.00

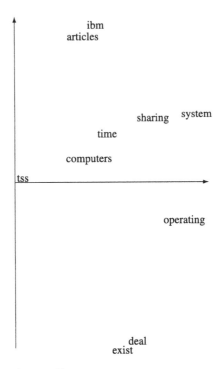

Fig. 3.15 Semantic space in a coordinate system

Each column is a description of the possible dimensions in which the context is given by the test collection and the query. The sign of the real number in an eigenvector indicates whether the corresponding word plays a positive role or a negative role; for example, the first column corresponds to the first eigenvector and tells us that every word plays a positive role in the dimension characterized by the term "operating system" since the top-weighted words are "operating" and "system." The second column refers to a dimension characterized by "articles" and "ibm," whereas two verbs ("deal" and "exist") are not participating in the implementation of this dimension since their sign is negative. If the first two columns are axes of a coordinate system, Fig. 3.15 shows the contextual dimensions from a different point of view; the words are placed as points, and the closer two words are to each other, the more these words describe a context.

Since the column vectors have length 1, they are suited to implementing both state vectors and projectors. If a column vector on the contrary implements a projector of an outcome of an observable, it may be used to measure the probability of the outcome. If a column vector implements a state vector, it defines a distribution of probability of the outcome of observables measured on the semantic space given by the test collection and the query. The column vectors can be combined to define either superposed states or mixed states; for example, the following vector:

$$a_1 \begin{pmatrix} 0.163 \\ 0.159 \\ 0.349 \\ 0.299 \\ 0.225 \\ 0.460 \\ 0.378 \\ 0.516 \\ 0.256 \\ 0.006 \end{pmatrix} + a_2 \begin{pmatrix} 0.429 \\ 0.061 \\ -0.49 \\ -0.51 \\ 0.444 \\ -0.12 \\ 0.183 \\ 0.195 \\ 0.135 \\ 0.003 \end{pmatrix}$$

is a superposed state vector where the a_is are amplitude. In contrast, a mixed state is

$$|a_1|^2 |v_1\rangle\langle v_1| + |a_2|^2 |v_2\rangle\langle v_2|$$

where the v_is refer to the first two column vectors.

3.6.2 *Entanglement in Semantic Spaces*

Hou and Song (2009) and Hou et al. (2013) started from HAL to model word correlation using entanglement. Suppose a semantic space is defined over a vector space of which a basis includes a vector for each possible combination of the words; for example, if the semantic space consists of "apple," "banana," and "cherry," the basis includes eight vectors indexed by a triple of bits such that the first bit refers to "apple," the second bit refers to "banana," and the third bit refers to "cherry"; it follows that $|000\rangle$ is the basis vector of the event that no word is occurring, $|001\rangle$ means that only "cherry" is occurring, the occurrence of "apple" and "banana" and the absence of "cherry" are represented by $|110\rangle$, and so on.

Suppose a pure state vector is defined over the observable values, these values being the triples of bits corresponding to the presence/absence patterns of the words. When three words are considered (e.g., "apple," "banana," and "cherry"), the pure state vector can be written as

$$|\phi\rangle = a_{000}|000\rangle + a_{001}|001\rangle + a_{010}|010\rangle + a_{011}|011\rangle +$$
$$a_{100}|100\rangle + a_{101}|101\rangle + a_{110}|110\rangle + a_{111}|111\rangle$$

where

$$|a_{000}|^2 + |a_{001}|^2 + |a_{010}|^2 + |a_{011}|^2 + |a_{100}|^2 + |a_{101}|^2 + |a_{110}|^2 + |a_{111}|^2 = 1$$

When the vector space of the basis vectors forming the pure state vector is bidimensional, entanglement means that the pure state vector cannot be expressed as a tensor product of two one-dimensional basis vectors. Algebraically, entanglement of a bidimensional pure state vector means that

$$a_{00}|00\rangle + a_{01}|01\rangle + a_{10}|10\rangle + a_{11}|11\rangle \neq (a_0'|0'\rangle + a_1'|1'\rangle) \otimes (a_0''|0''\rangle + a_1''|1''\rangle)$$

where the superscripts of the right-hand side refers to the position of the bit in the string of the basis vectors of the left-hand side. When the vector space of the state vector is more than bidimensional, the number of components of a tensor product is higher. In particular, when $|\phi\rangle$ is the state vector, a tensor product may be between two entangled state vectors as follows:

$$|\phi\rangle = (a_0|0\rangle + a_1|1\rangle) \otimes (b_{10}|10\rangle + b_{11}|11\rangle)$$

but

$$b_{00}|00\rangle + b_{01}|01\rangle \neq (b_0'|0'\rangle + b_1'|1'\rangle) \otimes (b_0''|0''\rangle + b_1''|1''\rangle)$$

or another tensor product may be between three state vectors as follows:

$$|\phi\rangle = (a_0|0\rangle + a_1|1\rangle) \otimes (b_0|0\rangle + b_1|1\rangle) \otimes (c_0|0\rangle + c_1|1\rangle)$$

As explained in Sect. 2.6, an entangled pure state vector can be prepared so that it cannot be rewritten as a tensor product. However, it is possible that $|\phi\rangle$ cannot be written as a tensor product of three bidimensional state vectors, but it can be written as a tensor product between a bidimensional state vector and a four-dimensional state vector; for example, $|\phi\rangle = \frac{1}{\sqrt{2}}|000\rangle + \frac{1}{\sqrt{2}}|111\rangle$.

3.7 Contextual Search

IR is intrinsically context dependent since what is relevant to one user in one place at one time can no longer be relevant to another user in another place or at another time. It follows that an IR system should be context aware. Search engines have been capturing some search environment features exploited at retrieval time; examples are location or search history as surveyed by Melucci (2012b) who earlier explained in 2008a how context can be viewed as a space of contextual dimensions which correspond to observables in QM. Consider two contextual dimensions, the user and the meaning of the term "bank" as depicted in Fig. 3.16. An observable measures the meaning of "bank" and has two basis vectors $|b_1\rangle$ and $|b_2\rangle$; another observable measures the user and has two basis vectors $|u_1\rangle$ and $|u_2\rangle$. The vectors $|b_1\rangle$ and $|b_2\rangle$ are mutually orthogonal as they refer to mutually exclusive contextual dimensions; the same applies to $|u_1\rangle$ and $|u_2\rangle$.

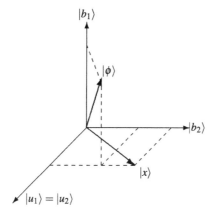

Fig. 3.16 Modeling context using subspaces

A context can be represented by a subspace spanned by basis vectors of the observable; for example, a context can be represented by the subspace spanned by $|b_1\rangle$ and $|u_1\rangle$, while another context can be represented by the subspace spanned by $|b_2\rangle$ and $|u_2\rangle$. Therefore, a context is represented by a subspace. The vectors $|b_1\rangle$ and $|u_1\rangle$ are mutually independent and form a basis of the subspace spanned by them; they are not necessarily orthogonal to each other since they refer to distinct observables. The same applies to the context represented by the subspace spanned by $|b_2\rangle$ and $|u_2\rangle$. A subspace includes all the vectors generated by a basis of that subspace; for example, the vector $|x\rangle$ in Fig. 3.16 belongs to the subspace spanned by $|u_2\rangle$ and $|b_2\rangle$. In general, the subspaces generated by the linear combination of basis vectors can also describe documents, queries, or other information objects.

To decide whether a dimension exists in an object, a measure of probability is needed. Therefore, the question is what is the probability that a contextual dimension is observed given an object represented by the state vector $|\phi\rangle$? For example, if $|\phi\rangle$ describes a document, what is the probability that the said document has been generated in the context corresponding to $|b_1\rangle$ and $|u_1\rangle$? The solution to this question would enable documents to be ranked with respect to a query by taking context into account. Thanks to the Gleason theorem, a state vector $|\phi\rangle$ simultaneously assigns probabilities to all questions involving contextual dimensions. The way the probabilities are assigned causes these probabilities to vary when object y varies, that is, the distribution of probability is dependent on the object y. As a consequence, two objects give different probabilities to one question involving some contextual dimensions. These two objects can then be ranked, thus answering questions like "In which object is a dimension i more likely to occur?" with the most probable object being presented first.

The use of vector bases to represent the context that has been exemplified above can be viewed as a departure from the VSM. The VSM assumes that there is a unique basis and that every vector is generated by that basis. Moreover, that basis spans the whole space or the number of basis vectors is the number of dimensions of the vector space. In particular, the canonical basis is often assumed as the unique basis according to the VSM, and then a document or query vector is given by the set of weights. This view is consistent with the fact that an IR system based on the VSM cannot take advantage of any context clues since the input of indexing is a document or a query produced out of context. Therefore, vectors are indifferent to context in the VSM, i.e., they are generated and ranked in the same way independently of the context in which the objects are found.

In contrast, a vector basis may be the construct for representing context. The basic idea is that, first, a vector is generated by a basis just as documents or queries are generated within a context. Second, every vector can be generated by different bases, and they belong to infinite subspaces; this is consistent with the fact that every information object is generated within different contexts. Finally, and as a corollary, the subspace spanned by a basis contains all the vectors that describe documents or queries in the same context; in that subspace, the vectors are related to each other by a linear combination.

3.7.1 Context and Projectors

Suppose $\{|b_1\rangle, \ldots, |b_k\rangle\}$ is a basis of a k-dimension subspace defined over the complex field. The way the basis generates an information object vector is given by

$$|\phi\rangle = \sum_{i=1}^{k} p_i |b_i\rangle \tag{3.10}$$

Looking at Eq. (3.10), the linear combination of some basis vectors well reflects the idea that context influences the materialization of information objects by combining different factors; for example, a document may be materialized by combining different meanings or aspects, such as informative content, space, time, or search history; alternatively, the basis vectors can describe different meanings given to terms, where each basis vector corresponds to a different meaning. In this way, two physically equivalent documents or queries can have a different meaning because their constituent terms correspond to different basis vectors. In this context, the p_is are a measure of the degree to which the respective basis vector $|b_i\rangle$s are chosen to generate $|\phi\rangle$; a different choice of the basis yields numerically different vectors.

The basis vectors describing the contextual factors may be mutually oblique, thus describing some dependence among factors. In contrast, orthogonal basis vectors cannot describe any dependence. The choice between orthogonal and oblique basis vectors in general depends on the domain and on design considerations. If orthonormal vectors represent term meanings, they may not be an appropriate encoding of the relationships between word senses, which do not need to be mutually unrelated. The appropriateness of the use of orthonormal or oblique vectors depends on the interpretation given to these vectors; for example, an eigenvector that results from the SVD of a term correlation matrix may be viewed as "primitive concept," and therefore, orthogonality may be quite appropriate; if, on the other hand, the basis vectors are interpreted as overlapped "clusters" of words, then oblique vectors would be a better choice and another decomposition, such as Cholesky's decomposition, would be preferred.

3.7.2 Probability of Context

Modeling documents or queries as random events is common in IR. The presence of context does not eliminate uncertainty, but rather raises the problem of measuring the degree to which some contextual factors occur in a document or query vector. The lack of knowledge about context means that a contextual factor, which is represented as a subspace, is a random event; for example, it is unknown whether a query should match an object in a mathematical context or whether a document can answer a query issued in a touristic context. Therefore, what one needs is a probability measure associated with the event that some contextual factors occur in an object. Such a probability measure would provide a measure of the degree to which the object represented by, say, $|\phi\rangle$ has been generated in the context represented by the subspace spanned by a basis B. Such a probability can be computed following the dictates of QM, that is,

$$P(y \text{ is generated by } B) = \text{tr}(\mathbf{B}|\phi\rangle\langle\phi|) = \langle\phi|\mathbf{B}|\phi\rangle$$

where $\mathbf{B}$ is the projector of the subspace spanned by B; note that when $|\phi\rangle$ is a linear combination of B, the probability is 1 and when $|\phi\rangle$ is orthogonal to all the vectors of B, the probability is 0 and that

$$P(y \text{ is not generated by } B) = 1 - \text{tr}(\mathbf{B}|\phi\rangle\langle\phi|) = \text{tr}((1 - \mathbf{B})|\phi\rangle\langle\phi|)$$

where $1 - \mathbf{B}$ is the projector of the subspace orthogonal to the subspace represented by $\mathbf{B}$.

The use of the trace rule above to compute the probability of context is explained by the Gleason theorem illustrated in Sect. A.2.2. This model can help design interactive and iterative functions starting from a given context represented by B and then by $\mathbf{B}$, collecting some additional input in terms of basis vectors which are then added to B for obtaining a new basis B' and computing a new projector $\mathbf{B}'$ which can finally be used to update the probability of context of information objects represented by $|\phi\rangle$.

A special case is given by the subspace spanned by a set of relevant document vectors. In this case, one speaks about probability of relevance which may be viewed as a special case of the probability of context. The probability that a document represented by $|\phi\rangle$ is relevant is given by

$$\text{tr}(\mathbf{R}|\phi\rangle\langle\phi|) = \langle\phi|\mathbf{R}|\phi\rangle$$

where $\mathbf{R}$ is the projector of the subspace spanned by a set of relevant documents.

A connection with LSA introduced by Deerwester et al. (1990) exists here. LSA decomposes a document-term matrix into three matrices by SVD and aims at extracting the most "meaningful" factors along which document content is described. Intuitively, the factors would correspond to concepts or term groups. The most "meaningful" factors along which the term meaning is described are also extracted by using the same decomposition. The term co-occurrence matrix usually built by LSA is Hermitian and can be built so as to have unitary trace. In this way, the term co-occurrence matrix can be seen as a density matrix. Once a density matrix is available, a probability distribution of the subspaces, e.g., an information object vector, can be assigned. In particular, the probability of the context described by a subspace can be computed by using that density matrix. Since a density matrix is Hermitian, it can be decomposed as

$$\rho = \gamma_1^2 \mathbf{B}_1 + \cdots + \gamma_k^2 \mathbf{B}_k$$

where $0 \leq \gamma_i^2 \leq 1$ and the $\mathbf{B}_i$s are projectors to the subspace spanned by the document vectors.

As explained in Sect. 1.2, a projector is like the sentence "Is this contextual factor occurring in this object?" which is either true (eigenvalue 1) or false (eigenvalue 0). A density matrix, therefore, is a linear combination of as many yes/no questions as the distinct eigenvalues, and the probability that the answer is "yes" is provided by the geometry of the subspaces. In other words, a density matrix incorporates the information about the occurrence of some contextual factors in terms of questions whose answers are subject to probability measures. Moreover, the eigenvectors of ρ provide the maximum likelihood estimators of the projector of the subspace. In other words, a basis can be seen as the parameter of a probability distribution attached to a random variable. In statistics, sample data are drawn in order to estimate the unknown parameter. Similarly, a term co-occurrence matrix can be seen as the sample being drawn to estimate the basis and the projector to the subspace. If the sample is assumed to be drawn from a context, the eigenvectors of ρ are the most likely factors of that context, and the γ_i^2s are the probabilities that the corresponding projectors contribute to the construction of the context represented by ρ.

3.8   Ranking Principle

A quantum PRP was proposed by Zuccon et al. (2009). This principle is based on the analogy between the double-slit experiment illustrated in Sect. 2.5 and the scenario in which a user has to decide whether to stop searching or to continue browsing the next document of a ranked document list after browsing until the last visited document.

Following the analogy with the double-slit experiment, the photon corresponds to the user and a slit represents a document, the event of passing from a slit is viewed as the action of examining the list of retrieved documents, and the measurement performed on the photon when it passed through a slit corresponds to the measurement of the action performed by the user, i.e., either continue searching or stop searching as depicted in Fig. 3.17 which should be compared with Fig. 2.11.

3.8.1   Probability Ranking Principle

Although the process of visiting a retrieved document list is quite complex, an IR system usually utilizes a much simpler model and applies the PRP to decide the n-th document in the list.

The PRP dictates that the best retrieval performance measured in terms of expected recall is achieved when the n documents with the highest probability of relevance are inserted in a list of n retrieved documents. This selection is performed one document at a time, and the decision about one document is taken independently of the decisions taken about other documents. The underlying assumption is that a user would view one document at a time, and viewing a document is an action independent of viewing other documents as if the user were receiving one document at a time in a sequential manner.

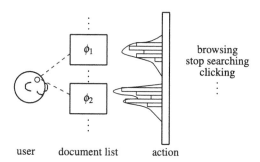

Fig. 3.17 The idea of the quantum PRP

According to the PRP, the expected recall of this strategy is the sum of the probability of relevance calculated for each retrieved document in the list, that is,

$$p_1 P(X = 1|\phi_1) + \cdots + p_n P(X = 1|\phi_n)$$

where ϕ_i is the event that the user is viewing the i-th document in the list with probability p_i and $X = 1$ is the event that the document is relevant (and the user may stop searching). Using the trace rule, we have that

$$P(X = 1|\phi_i) = p_i \text{tr}(|1\rangle\langle 1|\phi_i) = p_i \langle 1|\phi_i|1\rangle$$

where $|1\rangle\langle 1|$ is the projector of the event of relevance and ϕ_i is the density matrix of the state of the interaction step when the user is viewing the i-th document in the list. The expected recall is then based on the following mixture of density matrices:

$$p_1 \phi_i + \cdots + p_n \phi_n$$

This situation is similar to the one of the double-slit experiment where one slit is closed while the other slit is open. Similarly to when a photon passes through only one slit, the user decides to stop browsing, views only the n-th document, and ignores the others.

According to the PRP, the best decision taken by the IR system is proposing the n-th document with the highest probability of relevance to the user. The decision is independent of the documents already ranked and browsed by the user. Actually, the process of visiting a retrieved document list is much more complex than a straight list of links to the documents; a user may want to stop visiting, to go back to the previously visited documents, to skip the next retrieved documents or even pages of retrieved documents to arrive at a "random" document, or to type a Uniform Resource Locator (URL) directly in the toolbar; such a complexity was modeled partially by Brin and Page (1998) and further developed in the huge literature devoted to PageRank or devoted to other ranking principles such as Fuhr's (2008).

3.8.2   Quantum Probability Ranking Principle

In addition to the range of operations performed when visiting a list of retrieved documents, there is the situation in which a user is not only viewing the next document (or snippet), but he is also considering although unwillingly the documents thereafter. Suppose a user is browsing a ranked document list; this situation also comprises the event that the user just started to browse the list and no document has been visited yet. When the user has just viewed the $n - 1$-th document, he is going to view the n-th and decide whether to stop searching because the n-th document is deemed relevant.

The quantum PRP assumes that the user is viewing both documents and that the state of the interaction is a superposition; in particular, the state of viewing the n-th document is superposed on the state of viewing the $n-1$-th document. The state of the interaction is then described by the following state vector:

$$|\phi\rangle = a|\phi_n\rangle + b|\phi_{n-1}\rangle \qquad |a|^2 + |b|^2 = 1 \qquad \langle\phi_n|\phi_{n-1}\rangle = 0$$

where ϕ_n is the state vector of the interaction when the user is viewing the n-th document and a the amplitude of the probability that the user is viewing the n-th document. Therefore, instead of viewing only the n-th document, the user is simultaneously viewing two documents, i.e., both the n-th document and the $n-1$-th document. This situation is similar to the one of the double-slit experiment where both slits are open.

The probability that the user considers the n-th document relevant and stops searching can be computed for the superposed state. Suppose the user's action is described by an observable with the basis vectors $|0\rangle$ (to continue searching) and $|1\rangle$ (to stop searching at the current document). It follows that the probability of this event for the superposed state is

$$|\langle 1|\phi\rangle|^2 = |a|^2|\langle 1|\phi_n\rangle|^2 + |b|^2|\langle 1|\phi_{n-1}\rangle|^2 + I(\phi_n, \phi_{n-1})$$

where $I(\phi_n, \phi_{n-1})$ is the interference term between ϕ_n and ϕ_{n-1}. As the $n-1$-th document has already been viewed by the user, it is fixed for all the documents that remain to be ranked, and the principle states that the best decision is to select the document n that maximizes

$$|a|^2|\langle 1|\phi_n\rangle|^2 + I(\phi_n, \phi_{n-1})$$

The estimation of this quantity is not straightforward except for the first term. Indeed, $|a|^2$ can be estimated by some function decreasing with n as the Normalized Discounted Cumulative Gain (NDCG) function, for example. The probability $|\langle 1|\phi_n\rangle|^2$ can be estimated by the classical probabilistic models since it is the probability of relevance of the n-th document in the list. The problematic situation is given by the interference term. This term can be written as

$$2|a||b||\langle 1|\phi_n\rangle||\langle\phi_{n-1}|1\rangle|\cos\theta$$

where θ is the angle of the complex number

$$z = ab^*\langle 1|\phi_n\rangle\langle\phi_{n-1}|1\rangle$$

At first sight, the interference term is a real number which might be estimated after estimating the numbers multiplied in it. However, this is not the only solution since z is a complex number and there is an infinity of complex numbers z with a given $|z|$. As an alternative, the algebraic structure of z suggests that it describes some relationships between the observable of relevance and the state vectors of the documents viewed by the documents and between the documents thereof. However, it is still unclear, and open to further research, what these relationships are since the nature and the role played by the complex numbers need further investigation.

An approach to estimating the interference term may be based on estimating the moduli and then θ under the assumption that this angle may be like the angles encountered in the VSM (e.g., the angle between a query vector and a document vector). However, θ is part of $z = \cos\theta + i\sin\theta$. This is the reason why θ cannot be considered as the angle between two vectors as it might be suggested by the VSM because this θ is not referring to any angle, and the angles suggested by the VSM are defined within a vector space over the real field, whereas QM is defined within a vector space over the complex field.

3.9 User Interaction

The issues of user interaction in IR regard many aspects ranging from those extremely related to the cognitive level of the interaction to those related to the practical formulation and expansion of the user's queries. The application of the QM notions and formalism to the aspects of user interaction in IR has mainly concentrated on the algebraic representation of the notion of user's information need, information space, query, and other interaction features; the connections and the similarities with the research contributions to concept combination (Sect. 3.4), word ambiguity (Sect. 3.5), semantic spaces (Sect. 3.6), and contextual search (Sect. 3.7) should not come as a surprise.

There are three main streams of research on the application of the QM notions and formalism to the aspects of user interaction in IR: the combination of different contextual variables (e.g., interaction features and informative content features) within a common complex vector space to implement implicit feedback algorithms initiated by Melucci and White (2007), Melucci (2008b), and Melucci (2008a); the representation of information needs in a complex vector space and the use of the quantum mechanical operators to calculate the probability of relevance and the information need evolution initiated by Piwowarski and Lalmas (2009); and the representation of polyrepresentation in a complex vector space and the use of the quantum mechanical operators to describe complex information spaces initiated by Frommholz et al. (2010).

3.9.1 Implicit Relevance Feedback

The framework based on complex vector spaces illustrated in Sect. 3.7 is able to capture multiple aspects of user interaction using some notions of QM. The model uses display time, document retention, and interaction events to build a multifaceted user interest profile. Each facet of a profile can be represented by a vector which is usually an eigenvector. The set of the vectors representing the profile facets forms a basis of a vector space. Documents are matched against a user interest profile by computing a function of the distance between the document vector and the subspace spanned by the basis representing the user interest profile such that the closer the vector is to the subspace, the more the object is relevant to the profile.

To implement these vector spaces, the vectors that represent a user interest profile are computed by the SVD of the correlation matrix between the variables observed from a set of documents viewed by the user during the course of his search. The function of the distance between the document vector and the subspace spanned by the eigenvector is then used as a measure of the distance between the document and the profile. In the framework, the vectors that represent the profile are computed by the SVD of the correlation matrix between the variables observed from a set of documents seen by the user during the course of his search. As an example, suppose the following six variable (column) vectors have been observed after seeing six (row) documents:

$$
A = \begin{pmatrix}
\text{display time} & \text{scrolling} & \text{saving} & \text{bookmarking} & \text{access frequency} & \text{webpage depth} \\
1 & 0 & 3 & 7 & 6 & 7 \\
2 & 0 & 9 & 7 & 5 & 6 \\
2 & 0 & 7 & 6 & 4 & 5 \\
3 & 4 & 8 & 6 & 7 & 7 \\
4 & 1 & 3 & 6 & 5 & 5 \\
1 & 28 & 7 & 7 & 5 & 4
\end{pmatrix}
$$

where the depth of a webpage is the number of links from the root of the website to the webpage itself. The following variable correlation matrix is then computed:

$$
S = \begin{pmatrix}
+1.00 & -0.42 & -0.14 & -0.78 & +0.11 & +0.05 \\
-0.42 & +1.00 & +0.19 & +0.38 & -0.05 & -0.62 \\
-0.14 & +0.19 & +1.00 & +0.07 & -0.03 & -0.04 \\
-0.78 & +0.38 & +0.07 & +1.00 & +0.00 & +0.00 \\
+0.11 & -0.05 & -0.03 & +0.00 & +1.00 & +0.75 \\
+0.05 & -0.62 & -0.04 & +0.00 & +0.75 & +1.00
\end{pmatrix}
$$

The values of an eigenvector of S are scalars between -1 and $+1$; the further a value is from 0, the more the corresponding value is a significant descriptor of the profile facet represented by the eigenvector. The value can be likened to an index term weight. The sign can express the contrast between variables and then the presence of subgroups of variables in the same facet; for example, the first eigenvector is

$$|b_1\rangle = \begin{pmatrix} -0.479 \\ +0.516 \\ +0.170 \\ +0.436 \\ -0.308 \\ -0.436 \end{pmatrix}$$

and it tells us that saving is of little importance, since $|b_{13}| = 0.170$ is relatively close to zero, while the most important variables tend to cluster: scrolling and bookmarking tend to be performed together ($b_{12} = 0.516, b_{14} = 0.436$) and tend not to be performed when display time, access frequency, and browsing ($b_{11} = -0.479, b_{14} = -0.308, b_{16} = -0.436$) increase. Let $|y\rangle$ be an unseen document. The function of the distance between the document vector and the subspace spanned by the eigenvector is then used as a measure of the distance between the document and the profile facet. Therefore, the probability that y is pertinent to the facet represented by $|b_1\rangle$ is $|\langle y|b_1\rangle|^2$; for example, if the unseen document vector is, say,

$$|y\rangle = \begin{pmatrix} +0.71 \\ 0 \\ 0 \\ 0 \\ +0.71 \\ 0 \end{pmatrix}$$

then the probability is about 0.09.

3.9.2 Entangling Relevance and Behavior

The interaction involving a user which accesses a document for assessing the relevance for the task he is performing can be depicted in Fig. 3.18. The whole user-document interaction can be modeled as a composite system made of one document and one user. The system that results from the composition of one document and one user can fully be described by the state vector $|\phi\rangle$. It is of course a reduction of the representation that one should define if the real interaction setting were taken into account. However, the mathematical modeling of the user-system interaction is a very complex, if not prohibitive, task. The decision of modeling the interaction as a quantum composite system, which is in turn described as a state of the product space, comes from the objective of modeling Implicit Relevance Feedback (IRF) by using the quantum mechanical framework.

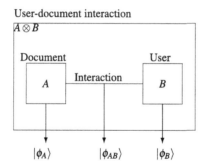

Fig. 3.18 Illustration of the interaction between a user and a document

The overall interaction system represented by the state vector $|\phi\rangle$ consists of two subsystems A and B. The system A refers to the visited document, whereas B refers to the user. A document is subjected to some observables such as the data about the visits paid to the document. Similarly, a user is subjected to some observables such as the relevance assessments made by him about the document. The document A and the user B are described by their respective state vectors $|\phi_A\rangle$ and $|\phi_B\rangle$. The system A and, consequently, the state $|\phi_A\rangle$ are of course an abstraction of the visited document under the assumption that the visit of a document can be reduced to a state vector. Similarly, the system B and, consequently, the state $|\phi_B\rangle$ are an abstraction of the user who visited the document.

The state vector $|\phi\rangle$ is an abstraction of the interaction between the user and the document. When separable, $|\phi\rangle$ is the tensor product $|\phi_A\rangle \otimes |\phi_B\rangle$. Separability means that the observation of one variable in A is uncorrelated with the observation of another variable in B. However, $|\phi\rangle$ may be entangled, thus witnessing a relationship between A and B. Under the hypothesis of entanglement, if one looks this interaction within the quantum mechanical framework, this hypothesis means that the states of the interaction between the user and document exist without being decomposed into distinct states $|\phi_A\rangle$ and $|\phi_B\rangle$. This means that the interaction may in principle be described by an entangled state $|\phi\rangle$ that cannot be explained in terms of $|\phi_A\rangle$ and $|\phi_B\rangle$.

The theory of entanglement and of density matrices suggests that the relationship between the behavior of the user when assessing document relevance and the way the document is visited can also be explained by the uncertainty about the preparation of the states of A and B. Although the connection might be exploited for IR purposes in further investigation, it seems that the behavior of the user when assessing the document relevance can be related to the style of interaction with the visited document; here, the style of interaction is condensed in a vector of values which measure the quantity of each feature, e.g., display, observed during interaction; the style is not therefore referred to the user, but rather to what one can observe about the visit of a document.

The formalism adopted in this section and the hypothesis that entanglement occurs in the user-document interaction recall the notion of collapse. When a user is visiting a document, the state of the system collapses to the pure state $|i_A\rangle$ of the observable that describes the measurement of the style of interaction. This happens because the user actually adopts a style of interaction when he accesses the document. When, for example, the user frequently clicks on the vertical scroll bar, the superposition of the basis states $|i_A\rangle$s collapses onto one of these basis states. After the state has collapsed, the coordinates of the other features take the values of the pure state onto which the original state collapsed. If the style of interaction is an observable, the collapse happens when this observable is measured. Similarly, when a user is assessing the relevance of a document, the state of the respective system B collapses to the pure state $|i_B\rangle$ of the observable that describes the measurement of relevance. This happens because the user actually decides that the document is relevant or not. What if the composite state of $A \otimes B$ is entangled? How does the state of B behave? What pure state of B does it collapse to?

The latter question is of interest in the domain of IR and in particular of IRF. Collapse of an entangled state means that the measurement of, say, A also induces a collapse of the state of B and vice versa. When the state of $A \otimes B$ is separable into two distinct states, the measurement of a property does not influence the state of the other observable. In other words, when a hypothetical IR system equipped with an IRF device wonders if a document is relevant, the state of the observable corresponding to the style of interaction collapses; for example, if

$$|\phi\rangle = \frac{1}{\sqrt{2}}|1_A\rangle \otimes |1_B\rangle + \frac{1}{\sqrt{2}}|2_A\rangle \otimes |2_B\rangle$$

is the state before the measurement of the style of interaction, which yields 1_A, then the state is $|1_A\rangle \otimes |1_B\rangle$ after the measurement, that is, the state of B collapses to $|1_B\rangle$. What is unclear is *when* the collapse occurs, given that it occurs. This is a fundamental question, and were these questions to be answered, the representation of the user and the visited document provided in this section would be a useful language for understanding how the interaction between user and system can be leveraged for improving retrieval effectiveness.

3.9.3 Information Need Representation

In this framework, the user's Information Need (IN) can be represented as a particle (e.g., a photon) subject to measurement through a suite of observables such as document relevance. It is also assumed that the user's IN is described by a pure state vector, i.e., as a unit vector in a complex vector space and that this state evolves while the user is interacting with the IR system. According to the QM framework, this pure state vector determines a probability distribution over the different observables; for example, there might be a relevance observable that provides the probability of relevance given a pure state vector representing an IN.

However, at the very beginning of a search process, an IN is not pure because it cannot be described precisely; if it were, it would be possible to meet the need retrieving all and only the relevant documents; the state is rather a mixture of all possible pure needs; therefore, it should be described by a mixed density matrix, and one can only know that the user is in one of all the possible pure vector states with a given probability.

The notion of superposition is crucial in describing how the INs are represented within the quantum mechanical framework. Suppose the vector space in which the pure state vector of an IN lies is spanned by a basis of two vectors. In particular, suppose that were the user asked to express the building blocks of his needs, he would list two words, i.e., "coffee" and "island." To these two words, two basis vectors can be defined, that is,

$$|\text{coffee}\rangle = \begin{pmatrix} 1 \\ 0 \end{pmatrix} \qquad |\text{island}\rangle = \begin{pmatrix} 0 \\ 1 \end{pmatrix}$$

These two vectors are mutually orthogonal[5] since they express two mutually occurring events, that is, were the user asked to express a word, he would express one word at a time, either "coffee" or "island." In order to represent a user looking for "java," it is assumed that this can be represented by (the projector associated to) the pure state vector

$$|\text{java}\rangle = \frac{1}{\sqrt{2}}|\text{coffee}\rangle + \frac{1}{\sqrt{2}}|\text{island}\rangle = \frac{1}{\sqrt{2}}\begin{pmatrix} 1 \\ 1 \end{pmatrix} \tag{3.11}$$

which is a superposition representing an IN different from the need represented by the vectors $|\text{coffee}\rangle$ and $|\text{island}\rangle$.

Alternatively, a mixture can be used to say that the user is interested in either "coffee" or "island" and that "java" should be intended as either the informal way to mean "coffee" or the large island in the Malay Archipelago. When a mixture is used, we have a mixed density matrix as follows:

$$\mu = \frac{1}{2}\begin{pmatrix} 1 & 0 \\ 0 & 0 \end{pmatrix} + \frac{1}{2}\begin{pmatrix} 0 & 0 \\ 0 & 1 \end{pmatrix} \tag{3.12}$$

which is to be interpreted by saying that with the probability that one half of the IN is about coffee and with the probability that one half is about the island.

---

[5]It should be recalled that the vectors of a basis might not be orthogonal since independence is the only requirement for a basis.

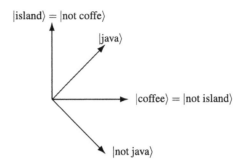

Fig. 3.19 Polyrepresentation, superposition, and mixture

Suppose there are three observables; each observable gives two mutually exclusive outcomes, that is, a document either is or is not about a topic. These three observables refer to "coffee," "island," and "java" and consist of the basis vectors |coffee⟩, |island⟩, and |java⟩ and the corresponding orthogonal vectors. Therefore, there are the following three pairs of basis vectors where a row refers to an observable as depicted in Fig. 3.19:

$$|\text{coffee}\rangle = \begin{pmatrix} 1 \\ 0 \end{pmatrix} \qquad\qquad |\text{not coffee}\rangle = \begin{pmatrix} 0 \\ 1 \end{pmatrix}$$

$$|\text{island}\rangle = \begin{pmatrix} 0 \\ 1 \end{pmatrix} \qquad\qquad |\text{not island}\rangle = \begin{pmatrix} 1 \\ 0 \end{pmatrix}$$

$$|\text{java}\rangle = \frac{1}{\sqrt{2}} \begin{pmatrix} 1 \\ 1 \end{pmatrix} \qquad\qquad |\text{not java}\rangle = \frac{1}{\sqrt{2}} \begin{pmatrix} 1 \\ -1 \end{pmatrix}$$

When the user's IN is described by the mixed state (3.12), the probability that a document is about "coffee" is

$$\text{tr}(\mu|\text{coffe}\rangle\langle\text{coffee}|) = \langle\text{coffee}|\mu|\text{coffee}\rangle = \frac{1}{2}$$

as expected since the IN is either "coffee" or "island" with uniform probability; therefore, the probability that a document is about "island" is again $\frac{1}{2}$. The probability that a document is about "java" is still $\frac{1}{2}$, that is, being

$$\text{tr}(\mu|\text{java}\rangle\langle\text{java}|) = \langle\text{java}|\mu|\text{java}\rangle = \frac{1}{2}$$

Since the superposition (3.11) is not representing a disjunction between "coffee" and "island," it is representing a new concept stemming from the superposition of two distinct concepts represented by $|\text{coffee}\rangle$ and $|\text{island}\rangle$. The calculation of the latter probability can indeed be stepped as follows:

$$\langle \text{java}|\mu|\text{java}\rangle = \left(\frac{1}{\sqrt{2}}\langle\text{coffee}| + \frac{1}{\sqrt{2}}\langle\text{island}|\right)\mu\left(\frac{1}{\sqrt{2}}|\text{coffee}\rangle + \frac{1}{\sqrt{2}}|\text{island}\rangle\right)$$

$$= \left(\frac{1}{\sqrt{2}}\langle\text{coffee}|\mu + \frac{1}{\sqrt{2}}\langle\text{island}|\mu\right)\left(\frac{1}{\sqrt{2}}|\text{coffee}\rangle + \frac{1}{\sqrt{2}}|\text{island}\rangle\right)$$

Since

$$\mu = \frac{1}{2}|\text{coffee}\rangle\langle\text{coffee}| + \frac{1}{2}|\text{island}\rangle\langle\text{island}|$$

we have that

$$\frac{1}{\sqrt{2}}\langle\text{coffee}|\mu = \frac{1}{2}\frac{1}{\sqrt{2}}\langle\text{coffee}|\text{coffee}\rangle\langle\text{coffee}| + \frac{1}{2}\frac{1}{\sqrt{2}}\langle\text{coffee}|\text{island}\rangle\langle\text{island}|$$

that is,

$$\frac{1}{\sqrt{2}}\langle\text{coffee}|\mu = \frac{1}{2}\frac{1}{\sqrt{2}}\langle\text{coffee}|$$

and

$$\frac{1}{\sqrt{2}}\langle\text{island}|\mu = \frac{1}{2}\frac{1}{\sqrt{2}}\langle\text{island}|$$

It follows that

$$\langle\text{java}|\mu|\text{java}\rangle = \left(\frac{1}{2}\frac{1}{\sqrt{2}}\langle\text{coffee}| + \frac{1}{2}\frac{1}{\sqrt{2}}\langle\text{island}|\right)\left(\frac{1}{\sqrt{2}}|\text{coffee}\rangle + \frac{1}{\sqrt{2}}|\text{island}\rangle\right)$$

$$= \frac{1}{4}\langle\text{coffee}|\text{coffee}\rangle + \frac{1}{4}\langle\text{island}|\text{island}\rangle$$

$$= \frac{1}{4} + \frac{1}{4}$$

$$= \frac{1}{2}$$

When the user's IN is described by the pure state (3.11), the probability that a document is about "coffee" is still

$$\text{tr}(|\text{java}\rangle\langle\text{java}|\text{coffee}\rangle\langle\text{coffee}|) = |\langle\text{java}|\text{coffee}\rangle|^2 = \frac{1}{2}$$

which is equal to the probability that a document is about "island," but not to the probability that a document is about "java," the latter being

$$\text{tr}(|\text{java}\rangle \langle \text{java}|\text{java}\rangle \langle \text{java}|) = |\langle \text{java}|\text{java}\rangle|^2 = 1$$

Therefore, as explained by Hughes (1989), the superposition (3.11) is distinguished from the mixture (3.12) not by the probabilities it assigns to the "coffee" and "island" observables but by those assigned to the "java" observable. It is the impossibility of expressing the "java" observable in terms of the "coffee" and "island" observables that enables us to distinguish between superposition and mixture. The graphical illustration of these vectors in a bidimensional plane shown in Fig. 3.19 highlights that the fact that different probabilities are assigned to the "java" observable is associated with the fact that the subspaces representing that observable are obliquely inclined to those representing the "coffee" and "island" observables.

So far, the difference between this model of INs and what the VSM provides is not apparent. An IN like the one symbolized by "java" and its superposition |java⟩ might be managed by an IR system based on the VSM by a query vector spanned by two basis vectors |coffee⟩ and |island⟩ (see also Sect. 1.3); therefore, if both the query vector and the document vector are (3.11), the score given by the system to the document is maximal. Actually, the difference is due to interference. Suppose a document is represented by

$$|x\rangle = a|\text{coffee}\rangle + b|\text{island}\rangle \qquad |a|^2 + |b|^2 = 1$$

If the inner product function between this vector and (3.11) of the VSM were utilized, the score assigned to the document would be $\frac{a+b}{\sqrt{2}}$. Besides, this is a complex number, which would make ranking impossible; the other difference is that the trace rule of the quantum mechanical frameworks would give

$$|\langle x|\text{java}\rangle|^2 = \frac{|a|^2 + |b|^2}{2} + I$$

where the last term of the right side is an interference term.

3.9.4 Document Polyrepresentation

Polyrepresentation consists of representing document content using different, possibly overlapping sources of evidence under the assumption that the documents that lie in this overlap are assumed to be relevant given a user's information need as reported by Ingwersen and Järvelin (2005). The principle that underlies polyrepresentation is that two documents that are similar to each other with respect to different representations tend to be relevant to the same information need.

The rationale of polyrepresentation is that when users seek information, the relevance assessment about a document depends on different dimensions and in general on more dimensions than the usual topical relevance used by IR systems. It is therefore not only a matter of how many query terms are contained in a document but also of the appropriateness of a document to the user's context as addressed in Sect. 3.7. Indeed, besides document topical content, the user's relevance assessment about a document involves different contextual aspects of a document such as query intent, personal interests, and document quality which interplay to determine whether a document contains information relevant to the user's information need, according to Melucci (2012b).

Polyrepresentation was not developed in a mathematical framework encompassing it completely and precisely for some years until it was described within the framework of QM. Such a framework should describe various aspects of document representation: fusion of document content representations, temporal aspects and dynamic changes, document structure and layout, and the relationships between these aspects according to Frommholz et al. (2010).

The quantum mechanical framework appears to provide such a mathematical language of polyrepresentation. The key idea is to define a series of bases of a complex vector space where each basis refers to an observable applicable to documents and INs; for example, a set of textual passages can be in correspondence with a vector basis, the list of authors of a library can be in correspondence with another vector basis, and so on for other observables other than index terms, ratings, reviews, etc. If these bases lie in the same vector space, they can be combined by linear combination so that each vector of a basis can be a liner combination of the vectors of another basis. If these bases lie in different spaces, they can be combined by tensor products so that additional larger bases can be constructed by tensoring smaller bases together; for example, given a basis about authors (e.g., $|\text{smith}\rangle$ and $|\text{miller}\rangle$) and a basis about terms (e.g., $|\text{apple}\rangle$, $|\text{banana}\rangle$, and $|\text{cherry}\rangle$), an additional larger basis can be constructed by tensoring the smaller ones together, thus obtaining $|\text{apple, smith}\rangle$, $|\text{banana, smith}\rangle$, $|\text{cherry, smith}\rangle$, $|\text{apple, miller}\rangle$, $|\text{banana, miller}\rangle$, and $|\text{cherry, miller}\rangle$. Documents and INs can then be expressed using these larger bases, and the trace rule can be applied accordingly.

3.10 Relevance Detection

Chapter 1 explained that the probabilistic models used to design and implement current systems are basically based on the theory for which events and probability distributions are represented, respectively, as sets and set measures which are in accord with Kolmogorov (1956)'s axioms; for example, the retrieval results are subsets of documents indexed by a term, indexed by a term and another term, not indexed by the term, and so on. Some measures of probability of relevance are applied to the document subsets. Hence, given some terms, the quest is how to optimally rank subsets of documents by some measure of relevance. Given that the probabilities of relevance of each document are as accurately estimated as possible, document ranking is optimized with respect to recall if the documents with the highest precision and fallout not greater than a given level are retrieved—the latter is the PRP stated in the paper by Robertson (1977).

3.10.1   Ranking Using Subsets

In Sect. 2.7, we introduced how detection can be viewed in QM. In this section, we utilize detection as an approach to modeling relevance detection and document ranking. Consider the subset of the documents indexed by a given term. Algebraically, this subset corresponds to, for example, $|1\rangle$ and $|1\rangle\langle1|$ which are, respectively, the vector spanning and the projector of the subspace corresponding to the subset of the documents indexed by the term; the complement subset can be represented by $|0\rangle$ and $|0\rangle\langle0|$.

Consider a situation in which documents are emitted by a source and set to a given pure state of relevance (i.e., relevant or not relevant), transmitted through a channel, and received by a detector which has to decide whether the state of the document (i.e., relevant or nonrelevant) is using symbols such as index terms; Fig. 3.20 depicts this setting.

The index term frequencies observed can serve to decide about relevance. The decision taken depends on the range of frequencies to which an observed frequency belongs. If the observed frequency belongs to a certain region of acceptance, the retrieval system decides that a document was, say, relevant; otherwise, it was not relevant. If more index terms are used to decide about relevance as is customary when a query is submitted by a user, the IR system computes a score for each document according to a model; the score is then matched against a region of acceptance used to retrieve or rank the documents.

In IR, the pure state vectors $|\phi_0\rangle$ and $|\phi_1\rangle$ might, respectively, correspond to "nonrelevance" and "relevance," whereas x refers to the outcome of a binary variable describing the occurrence of a given index term; one can conceive other states such as aboutness, authoritativeness, or other similar properties. Usually, the symbols used to encode the occurrence of a term are 0 and 1, thus obtaining

$$P(\text{term occurs in a relevant document}) = |\langle1|\phi_1\rangle|^2$$

Fig. 3.20 IR as a detection problem

and

$$P(\text{term occurs in a nonrelevant document}) = |\langle 1|\phi_0\rangle|^2$$

In IR, the following simple setting can be prepared. Each document is indexed by only one term x such that the observable can yield either 0 or 1. A document may be either relevant (i.e., in state ϕ_1) or nonrelevant (i.e., in state ϕ_0). Suppose it is decided to retrieve a document when the term occurs; otherwise, the document is not retrieved; retrieving a document means that the decision is that the document is relevant. Algebraically, we have that $A_0 = \{0\}, A_1 = \{1\}$ and then that

$$\mathbf{A_0} = |0\rangle\langle 0| \qquad \mathbf{A_1} = |1\rangle\langle 1|$$

Suppose that relevance is described by two pure state vectors $|\phi_0\rangle$ and $|\phi_1\rangle$. It follows that

$$Q_d = q_0|\langle 0|\phi_0\rangle|^2 + q_1|\langle 1|\phi_1\rangle|^2$$

As another example, suppose an IR decides to always retrieve the document, that is, $A_0 = \emptyset, A_1 = \{0, 1\}$. It follows that

$$\mathbf{A_0} = \mathbf{0} \qquad \mathbf{A_1} = |0\rangle\langle 0| + |1\rangle\langle 1| = \mathbf{1}$$

and that

$$Q_d = q_0\text{tr}(\mathbf{0}|\phi_0\rangle\langle\phi_0|) + q_1\text{tr}(\mathbf{1}|\phi_1\rangle\langle\phi_1|) = q_1\text{tr}(|\phi_1\rangle\langle\phi_1|) = q_1$$

that is, the decision is the probability of correct when the document is relevant a priori.

When the density matrices representing the states of a particle are mixtures of pure density matrices, the optimal projectors have a special form. Suppose a collection of documents can be partitioned in $k = 2^d$ subsets corresponding to the possible combination of the occurrence of d index terms; for example, when $d = 2$ index terms are considered, $k = 4$ document subsets can be built, i.e., the subset of documents not indexed by either term represented by the basis vector $|00\rangle$; the subset of documents indexed by the first term, not by the second term, and represented by the basis vector $|10\rangle$; the subset of documents indexed by the second term, not by the first term, and represented by the basis vector $|01\rangle$; and the subset of documents indexed by both terms and represented by the basis vector $|11\rangle$.

Each basis vector corresponds to a projector; when $d = 2$, we have four projectors:

$$\mathbf{C_0} = |00\rangle\langle 00| \qquad \mathbf{C_1} = |01\rangle\langle 01| \qquad \mathbf{C_2} = |10\rangle\langle 10| \qquad \mathbf{C_3} = |11\rangle\langle 11|$$

where

$$\mathbf{C}_0 + \mathbf{C}_1 + \mathbf{C}_2 + \mathbf{C}_3 = 1 \qquad \mathbf{C}_i\mathbf{C}_j = 0 \quad i \neq j$$

Consider the following mixed density matrices:

$$\rho_0 = p_{0,0}\mathbf{C}_0 + \cdots + p_{0,k-1}\mathbf{C}_{k-1}$$
$$\rho_1 = p_{1,0}\mathbf{C}_0 + \cdots + p_{1,k-1}\mathbf{C}_{k-1}$$

where $p_{j,s}$ is the probability that a document contains the terms according to $|s\rangle$, s being a string of 0s and 1s (e.g., $s = 10$) when the state is ϕ_j; for example, when $k = 2^2 = 4$, we have that

$$\rho_0 = p_{0,0}\mathbf{C}_0 + p_{0,1}\mathbf{C}_1 + p_{0,2}\mathbf{C}_2 + p_{0,3}\mathbf{C}_3$$
$$\rho_1 = p_{1,0}\mathbf{C}_0 + p_{1,1}\mathbf{C}_1 + p_{1,2}\mathbf{C}_2 + p_{1,3}\mathbf{C}_3$$

In general, it follows that the optimal projector of

$$q_1\rho_1 - q_0\rho_0 = (q_1p_{1,0} - q_0p_{0,0})\mathbf{C}_0 + \cdots (q_1p_{1,k-1} - q_0p_{0,k-1})\mathbf{C}_{k-1}$$

can be defined as

$$\mathbf{Q}_1 = \sum_{q_1p_{1,s}>q_0p_{0,s}} \mathbf{C}_s$$

In IR, this means that the documents represented by the string s such that the likelihood of relevance $(p_{1,s})$ is greater than the likelihood of nonrelevance $(p_{0,s})$ are retrieved. In actual application, an IR system not only decides whether to retrieve the documents represented by s on the basis of the likelihood of relevance; it also ranks the documents by the ratio between the likelihoods (i.e., the likelihood ratios) or by the difference between the likelihoods. Consider the following example. A document collection is indexed by one index term only. This term can occur in a relevant document with probability p_1 and can occur in a nonrelevant document with probability p_0. Occurrence can be represented by the pair of projectors

$$|0\rangle\langle 0| = \begin{pmatrix} 0 & 0 \\ 0 & 1 \end{pmatrix} \qquad |1\rangle\langle 1| = \begin{pmatrix} 1 & 0 \\ 0 & 0 \end{pmatrix}$$

and the relevance states are represented by the density matrices

$$\rho_0 = \begin{pmatrix} p_0 & 0 \\ 0 & 1 - p_0 \end{pmatrix} \qquad \rho_1 = \begin{pmatrix} p_1 & 0 \\ 0 & 1 - p_1 \end{pmatrix}$$

It can easily be checked that

$$p_0 = \text{tr}(\rho_0|1\rangle\langle1|) \qquad p_1 = \text{tr}(\rho_1|1\rangle\langle1|)$$

The problem is to find a partition of the set of observable values $\{0, 1\}$ such that the probability of correct decision is maximum. Consider the SVD of the matrix $q_1\rho_1 - q_0\rho_0$ which yields the following decomposition:

$$(p_1q_1 - p_0q_0) \begin{pmatrix} 1 & 0 \\ 0 & 0 \end{pmatrix} + ((1 - p_1)q_1 - (1 - p_0)q_0) \begin{pmatrix} 0 & 0 \\ 0 & 1 \end{pmatrix}$$

The solution depends on the eigenvalues and ultimately on p_0, p_1. The region of acceptance is given by the projectors whose eigenvalues are positive. Suppose $q_0 = q_1 = \frac{1}{2}$; we have that

$$p_1q_1 - p_0q_0 > 0 \quad \text{if and only if} \quad p_1 > p_0$$

and that

$$(1 - p_1)q_1 - (1 - p_0)q_0 > 0 \quad \text{if and only if} \quad p_1 < p_0$$

The eigenvalues cannot both be positive, and therefore, ϕ_1 cannot be accepted regardless of the value that is observed. Therefore, ϕ_1 is accepted when 1 is observed (i.e., the index term occurs) and $p_1 > p_0$. Note that this result is the same as the result of the classical BIR model illustrated, for example, by van Rijsbergen (1979). This example suggests that the projectors providing a solution to the maximization problem are a combination of the projectors of the observable corresponding to the terms used to encode the documents. However, this is not always the case as illustrated in the next section since there are more subspaces than subsets.

3.10.2 Ranking Using Subspaces

There are projectors for which a subset of documents does not exist since they are in between and oblique to the subspaces; an example is provided in Fig. 2.5 where both $|\eta_0\rangle$ and $|\eta_1\rangle$ span rays oblique to those spanned by $|0\rangle$ and $|1\rangle$. The calculation of $|\eta_0\rangle$ and $|\eta_1\rangle$ is provided in Sect. 2.7. Suppose that the subspaces that correspond to the subsets (i.e., $\emptyset$, $L(0)$, $L(1)$, or $L(0) \vee L(1)$ using the notation of Sect. 2.2) yielded by dint of the PRP are rotated to get $|\eta_0\rangle$ and $|\eta_1\rangle$. Computing these oblique eigenvectors is like rotating the observable vectors $|0\rangle, |1\rangle$ of Fig. 2.5 by a given angle. When the rotation angle is a multiple of $\pi/2$, the resulting vectors are orthogonal to the observable vectors. In this event, the resulting vectors correspond to projectors orthogonal to the projectors defined from the observable vectors. Since they are orthogonal, they represent compatible observables. When the observables are compatible, subspaces turn out to be equivalent to subsets and can then be subject to set operations (see Sect. 1.2 and also the book of Griffiths (2002)), and the rotation does nothing other than define alternative subsets of the space of the observable values and therefore define alternative regions of acceptance.

As the PRP states the best region of acceptance in terms of recall and fallout, the rotation to orthogonal subspaces only defines worse regions of acceptance than that stated by the PRP. This means that when the subspaces are equivalent to subsets, the sets that would correspond to the eigenvectors $|\eta_0\rangle$ and $|\eta_1\rangle$ can classically be reformulated in terms of the sets of documents corresponding to $|0\rangle$ and $|1\rangle$, and thus, the eigenvectors represent something different from the observable vectors, yet they can be interpreted as Boolean expressions of the propositions corresponding to the observable vectors. It may turn out that after a rotation by a multiple of $\pi/2$, the eigenvectors $|\eta_0\rangle$ and $|\eta_1\rangle$ that result from the rotation are the negation of the observable vectors $|0\rangle$ and $|1\rangle$, and therefore, the physical interpretation of the eigenvectors is straightforward; for example, after a rotation by exactly $\pi/2$, the eigenvectors $|\eta_0\rangle$ and $|\eta_1\rangle$ that result from the rotation are such that $|\eta_0\rangle = |0\rangle$ and $|\eta_1\rangle = |1\rangle$, and therefore, the physical interpretation of the eigenvectors is the same interpretation as the observable vectors.

When the angle is different from a multiple of $\pi/2$, $|\eta_0\rangle$ and $|\eta_1\rangle$ of the figure are obtained, and the subspaces are not equivalent to subsets. When the subspaces are not equivalent to subsets, the sets that would correspond to the property that corresponds to an observable vector cannot be observed. Although the PRP states the best region of acceptance in terms of recall and fallout when the density matrices are mixtures (see Sect. 2.7), the rotation to orthogonal subspaces defines a region of acceptance which is different from that stated by the PRP. This means that when the subspaces are not equivalent to subsets, the sets that would correspond to the eigenvectors $|\eta_0\rangle$ and $|\eta_1\rangle$ cannot classically be reformulated in terms of the sets of documents corresponding to $|0\rangle$ and $|1\rangle$, and thus, the eigenvectors cannot be interpreted as Boolean expressions of the propositions corresponding to the observable vectors. In general, the eigenvectors $|\eta_0\rangle$ and $|\eta_1\rangle$ cannot classically be reformulated in terms of $|0\rangle$ and $|1\rangle$, and thus, the eigenvectors cannot be easily interpreted. The fact that optimal projectors of "oblique" subspaces can be found is discussed in the next section.

3.10.3 Optimal Ranking

From a detection point of view, the question is whether the subspaces spanned by $|\eta_0\rangle$ and $|\eta_1\rangle$ can be more powerful than the subsets dictated by dint of the PRP. To answer this question, Melucci (2012a) wondered what would happen to ranking and then to the PRP if a document collection were represented using vector subspaces. It was found that oblique subspaces that can be more effective than the classical subsets of the PRP can be found if the probability distributions are provided by pure state vectors. Suppose $|\phi_0\rangle, |\phi_1\rangle$ are two pure state vectors. The region of acceptance that has a higher probability of detection at every probability of false alarm than the region of acceptance given by the PRP applied on sets is given by the eigenvectors $|\eta_0\rangle, |\eta_1\rangle$ of (2.12) which are depicted in the tridimensional space in Fig. 2.5. The probability of the correct decision Q_d calculated using η_0 and η_1 is never less than the probability of the correct decision P_c calculated when the region of acceptance is given by document subsets at every probability of false alarm. Therefore, ranking by document subspaces is in principle more effective than ranking by document subsets, and a new ranking principle can be stated to make retrieval more general than the PRP. These results raise some issues though.

The rotation that moves $|0\rangle$ and $|1\rangle$ to $|\eta_0\rangle, |\eta_1\rangle$ unfortunately does not correspond to a physical transformation of terms and document subsets to other sets. For now, we have to accept it as adequate despite wanting something more concrete and think about this transformation on quite an abstract level. This situation is known as incompatibility between observables in QM since $|\eta_0\rangle, |\eta_1\rangle$ are only two superpositions of the observable given by $|0\rangle$ and $|1\rangle$.

The impossibility of expressing $|\eta_0\rangle$ and $|\eta_1\rangle$ using "classical" document subsets points out the issue of the measurement of the observable vectors. Measurement means the actual finding of the presence/ absence of an observable via an instrument or device. The measurement of term occurrence is straightforward because occurrence is a physical property measured through an instrument or device (a program that reads texts and writes frequencies is sufficient). Using the classical probability theory, if we observe a word, we measure the probability that every document described by the word is either relevant or not relevant. This view promptly links to the view of the document collection as an ensemble in which the uncertainly is given by the distribution of the observable values in the ensemble.

In contrast, the measurement of what corresponds to $|\eta_0\rangle$ and $|\eta_1\rangle$ is much more difficult because no physical property corresponds to them and cannot be expressed in terms of word occurrence as outlined by Griffiths (2002, pages 54–64). Thus, the question is: what should we observe from a document so that the outcome corresponds to the eigenvector? Indeed, Q_e and Q_d can be achieved only if an IR system is capable of observing the eigenvectors $|\eta_0\rangle$ and $|\eta_1\rangle$. If an IR system observed $|\eta_0\rangle$ or $|\eta_1\rangle$ in a document, the system would decide whether the document is relevant with probability of error Q_e. In particular, if we were able to give an interpretation to the eigenvectors, retrieval and indexing algorithms could measure those vectors. The problem is that the view given by $|\eta_0\rangle$ and $|\eta_1\rangle$ links to the view of a document as a particle in a superposed state in which the uncertainty is given by the random collapse of the superposed state to one of the observable values.

Three interpretations of the eigenvectors $|\eta_0\rangle, |\eta_1\rangle$ can be provided. An interpretation is geometric because it is in terms of vectors. Another interpretation is probabilistic. Yet another interpretation is logical since the eigenvectors are expressions of propositions.

From a geometric point of view, each vector is a superposition of another two independent vectors. Figure 2.5 depicts the way the vectors interact in the tridimensional space, whereas Fig. 2.20a is a pictorial description in the bidimensional space. These figures show that the observation of a binary feature places the observer upon either $|0\rangle$ or $|1\rangle$ whenever he measures $|0\rangle$ or $|1\rangle$, respectively. The measurement of the observable corresponding to $|\eta_0\rangle$ and $|\eta_1\rangle$ should be performed by a device which is at angle with the device performing the measurement of the observable corresponding to the basis $\{|0\rangle, |1\rangle\}$ as depicted in Fig. 2.21b. However, superposition cannot correspond to an easy transformation in IR, and there is no way to move upon $|\eta_0\rangle$ or $|\eta_1\rangle$ because no measurement device behaving like the one in Fig. 2.21b is available yet.

From a probabilistic point of view, the eigenvectors and the state vectors are related by the trace rule. As these eigenvectors are mutually orthonormal by definition, they induce a valid probability distribution. Indeed they implement an observable, and the state vectors provide the probability distribution of this observable. The probability distribution may provide some constraint which must be held by $|\eta_0\rangle$ and $|\eta_1\rangle$ so that they represent the optimal observable dictated by the decomposition of (2.12); for example, the observable must be defined so that $Q_d = 1 - Q_e$. However, a probability distribution cannot provide us with a deterministic way to observe the features which correspond to the eigenvectors.

From a logical point of view, the eigenvectors are assertions like the observable vectors are, e.g., $X = x$ corresponds to $|x\rangle$. According to this view, the optimal eigenvectors $|\eta_0\rangle$ and $|\eta_1\rangle$ are assertions about the documents. As they are superpositions of $|0\rangle$ and $|1\rangle$, it is difficult to express them using the classical logic. Indeed, the logic used to combine subspaces is more general than the logic used to combine subsets, and what we express using subspaces might not be expressed using subsets because the distributive law fails.

3.10.4   Using Quantum Detection in Relevance Feedback

In this section, we illustrate how the RF algorithms may be inspired by the principles of quantum detection. In summary, these algorithms may build query vectors as the optimal detectors of a quantum signal detection system. These optimal detectors will have to decide the (unknown) relevance state of a document on the basis of the available data, e.g., query term frequency. Technically speaking, these algorithms project the original query vector on a special subspace which is given by the principles of quantum detection. The vector that results from the projection will be matched against the vectors of the documents by the inner product function described in Sect. 1.3.2.

More precisely, let $|\eta_1\rangle$ be the vector spanning the subspace given by the principles of quantum detection. The projection of the original query vector can be expressed as

$$|\eta_1\rangle\langle\eta_1|y\rangle$$

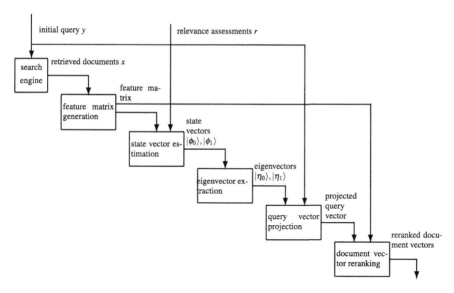

Fig. 3.21 General Relevance Feedback algorithm inspired by the principles of quantum detection

and it is essentially the query term re-weighting scheme proposed in this section. If $|x\rangle$ is a document vector, ranking is computed by

$$\langle x|\eta_1\rangle\langle\eta_1|y\rangle \tag{3.13}$$

The general RF algorithm inspired by the principles of quantum detection is depicted in Fig. 3.21. The initial query represented as a vector $y \in \mathbb{R}^d$ is input to the search engine of the IR system. The engine outputs a ranked list of N documents represented by the vectors x. The engine makes use of these document vectors for generating a $N \times d$ feature matrix to be used to estimate the state vectors $|\phi_0\rangle$ and $|\phi_1\rangle$ according to the relevance assessments r. Then, the matrices $|\phi_0\rangle\langle\phi_0|$ and $|\phi_1\rangle\langle\phi_1|$ are calculated, and the eigenvectors $|\eta_0\rangle$ and $|\eta_1\rangle$ are extracted. After the projection of the query vector on $|\eta_1\rangle$, the document vectors are reranked.

3.11 Evaluation Studies

In IR, the main object of evaluation is a system since it is of interest to the researchers to know whether the system under investigation is better than others. To evaluate a system, it is crucial to recognize the components of the system thereof because evaluation often concentrates on a few components, and in many cases, it does so on one component alone. If a system is regarded as a set (or something more structured than a set) of components, one speaks about system configuration, and what is subjected to evaluation is a configuration. Researchers therefore often compare configurations and struggle to obtain the best configuration in the field; for example, a configuration might be given by the most effective tuple of parameters of the BM25 weighting scheme. Although a component may be either hardware or software, evaluation is concerned with software components since these are the artifacts designed and implemented by the researchers in IR. Efficiency is extremely important in industrial search engines, but academic research is mostly concerned with effectiveness and therefore with the proportion of documents that are relevant to an information need and retrieved against a query representing this need. Despite the long history of experiments, the perfect IR system is still a chimera especially due to the intrinsic difficulty in ascertaining relevance in a very dynamic context.

Evaluation is carried out through a series of studies, i.e., a detailed investigation and analysis of a system configuration. According to the aims, a study can be descriptive to describe how the configuration works, explicative to explain how a variable interacts with other variables, or explorative to understand in depth how a specific configuration works. According to the environment where a study is performed, there are laboratory studies when the study is performed in controlled spaces and time spans relatively isolated from the external noise, and there are naturalistic studies when derived from or closely imitating real environments where the configuration is actually utilized by the end users. A study can also be a user study when it is performed, thanks to the participation of humans or when the main object of the evaluation is the user. When the same group of users is observed over a period of time, the study is called longitudinal. A study is a case study when it concentrates on a particular user, configuration, user group, or situation over a period of time.

In IR, studies are often based in a laboratory and utilize samples of the document collection to which the system configuration under investigation should be devoted. The main instrument used in the laboratory studies is the test collection obtained when the sample documents are combined with artificial representations of the users and of the relevance assessments. A test collection consists of three sets: a test document set, a test topic set, and a relevance assessment set. A test topic is a relatively short textual document about a subject used to formulate the queries given as input to the system configuration. A relevance assessment is basically a pair document-topic mapping to a relevance degree, e.g., a document is relevant to a topic or not.

The comparison between the system configurations under investigation requires the computation of measures and in particular of effectiveness measures; efficiency measures are mostly concerned with the hardware components or with the data structures and the algorithms employed to implement a configuration. Historically, the main measures are precision and recall. Recall is the proportion of relevant documents that are retrieved against a query. However, it is by now disused since it requires the knowledge of the complete set of relevant documents for each topic, although this knowledge would be crucial to understand the upper bound of an effectiveness measure. Precision is the proportion of retrieved documents that are relevant.

This section reports the main experimental results obtained since the quantum mechanical framework was proposed to design indexing and retrieval methods. Most evaluation studies carried out in this field were concerned with comparing some indexing or retrieval methods inspired by QM with the classical methods utilized within the current IR technology. The methods inspired by QM and experimented by the researchers of this field indeed utilized some quantum mechanical concepts (e.g., superposition and projectors) to represent the state or the observables of an IR object (e.g., a document or a query); these methods exploited this representation within an indexing or retrieval algorithm and were compared with the classical methods using test collections or user studies. Most of the studies are empirical investigation which suggested the potential of the quantum mechanical framework applied to IR.

3.11.1 Vector Negation and Ambiguity

Widdows (2003) performed the early experiments to assess the effectiveness of vector negation using some notions of the quantum mechanical framework illustrated in Sect. 3.5. The experiments aimed at measuring the capability of word negation in removing the documents containing the negated words from the retrieved document set without removing the documents containing the positive words. The word negation methods tested in the experiments were compared with the baseline alternative of no negation at all.

Note that the notion of effectiveness on which experiments on word negation and ambiguity were carried out was quite different from the notion usually utilized in IR. Indeed, no traditional test collection was employed and no document was labeled as relevant to an information need as it happens when test collection are employed in IR experimentation. The experiments carried out by Widdows were actually counting the number of occurrences of positive words and the number of occurrences of negative words under the assumption that a document was relevant to the meaning of an expression like "positive term AND NOT negative term" when it contained as many references to the positive term and as few references to the negative term as possible.

Given the assumption about relevance, with the purpose of obtaining reasonable results, the experiments were performed using document corpora which are quite homogeneous in content, written in one language (i.e., English) and presumably based on a relatively limited vocabulary. In particular, the British National Corpus by Oxford University, the New York Times News Syndicate by the Linguistic Data Consortium, and the Ohsumed corpus of medical documents of Hersh et al. (1994) were utilized. The measures reported in the evaluation study are the relative frequencies of a term in a set of documents; in particular, for each set of retrieved documents, the relative frequency of the positive term and the relative frequency of the negated term were reported. These measures aimed to give a degree to which a method of word negation was able to differentiate between documents indexed by a positive term and documents indexed by a negative term.

As a general comment, the assumption under which a document was considered relevant when it contained as many occurrences of the positive term and as few occurrences of the negative term as possible can undermine the correct interpretation of the experimental results from the perspective of IR effectiveness. When retrieval effectiveness is the focus of evaluation, it is crucial that the test collections that are utilized in the experiments incorporate the "true" relevant documents and that relevance is assessed with respect to the real user's information need. Indeed, when negative RF, which is the retrieval method closest to vector negation, was evaluated in IR, the experiments gave inconsistent results; some results were encouraging, while others testified that removing evidence from a query had often a detrimental effect on retrieval effectiveness as reported, for example, by Harman (1992). Further experiments would be useful by exploiting, for example, the narrative of the topics which indicate the characteristics of nonrelevant documents to implement the vector to be subtracted from the query vector.

3.11.2   Contextual Search

The experiments in contextual search reported by Melucci (2008a) were based on the estimation of a density matrix using some documents under the assumption that this set can be a source of evidence to estimate a context; a density matrix was used to mathematically describe a context.

A subspace describing a set of relevant documents and a subspace built from nonrelevant documents were described by one density matrix each. The probability that a document represented by $|y\rangle$ is relevant is given by $\mathrm{tr}(\rho_R |y\rangle\langle y|)$ where ρ_R is the density matrix estimated by the set of relevant documents. Similarly, the probability that a document represented by $|y\rangle$ is not relevant is given by $\mathrm{tr}(\rho_{\bar{R}} |y\rangle\langle y|)$ where $\rho_{\bar{R}}$ is the density matrix estimated by the set of nonrelevant documents.

Starting from a test collection, a topic was assigned a set of relevant documents and a set of nonrelevant documents explicitly assessed by the assessors or implicitly indicated by the system by Pseudo Relevance Feedback (PRF) or by the end user while interacting with the system. After removing stop words and stemming keywords, the two sets of documents, either relevant or nonrelevant, were processed to build up a number for each word which measures how often that word appeared within a window of text centered around the word. Windows are used for putting words into their own context—in the experiments, context is meant as the window around a word—and for relating the words to the others in the same context. In this way, the resulting co-occurrence matrix represents the data collected, its columns and rows refer to content bearing words, and each matrix element stores a similarity values.

The eigenvectors $|\phi_1\rangle$ and $|\phi_2\rangle$ of a co-occurrence matrix can be used as state vectors such that one state vector defines a density matrix. Each eigenvector can be viewed a dimension of relevance, i.e., a possible way the selected keywords interact when occurring in a set of relevant documents; the same can be applied to a co-occurrence matrix calculated from a set of nonrelevant documents. The density matrices can be linearly combined to obtain a mixed distribution as follows:

$$\rho_R = \lambda_1 |\phi_1\rangle\langle\phi_1| + \lambda_2 |\phi_2\rangle\langle\phi_2|$$

where the λs are the two largest eigenvalues yet might be replaced with two probabilities that measure the importance of each dimension of relevance represented by an eigenvector. The same procedure can be applied to obtain a mixed density matrix from nonrelevant documents. The probability that a document has been generated by the context of relevance can be computed using the mixed density matrix estimated from relevant documents; similarly, the probability that a document has been generated by the context of nonrelevance can be computed using the mixed density matrix estimated from nonrelevant documents. The probabilities computed in this way can be utilized to rank the documents. Figure 3.22 depicts the two principal eigenvectors estimated from the set of AP88 relevant documents of topic 255.

keyword	$|\phi_1\rangle$	$|\phi_2\rangle$
china	0.11338	0.15884
citi	0.17166	0.34414
countri	0.17592	−0.10948
environ	0.25244	0.33574
environment	0.28902	0.26769
factori	0.18635	0.01760
govern	0.28153	−0.05647
industri	0.26364	−0.04916
million	0.18622	−0.09272
nation	0.20032	−0.01514
offici	0.22023	0.21031
pollution	0.25666	−0.32608
problem	0.14833	−0.17257
protect	0.22541	0.16492
report	0.37851	−0.51922
river	0.15458	0.12263
shanghai	0.15561	−0.27667
wast	0.25982	0.08156
water	0.15099	0.25903
year	0.22871	0.05270

Fig. 3.22 Principal eigenvectors of a co-occurrence matrix estimated from relevant documents

3.11.3 Implicit Relevance Feedback

While Melucci (2008a) addressed contextual search using the quantum mechanical framework to model PRF, Melucci and White (2007) modeled the context observed and measured during the interaction between a user and a set of webpages. In particular, the context was given by the information about the documents seen by the user and by the features of the interaction. This information was then processed for extracting eigenvectors from which density matrices were computed to represent the contextual factors used to rerank documents through an IRF algorithm. The aim of the experiment was to compare the retrieval effectiveness of multiple IRF algorithms that used different sources of implicit feedback and translate this feedback into document rankings. To evaluate the performance, the interaction logs of real subjects were used to simulate a user who accesses a series of documents and performs some actions such as reading, scrolling, bookmarking, and saving. The IRF algorithms under investigation were assumed to be part of a system monitoring subject behavior and using these interaction data as a source of IRF to retrieve and order the unseen documents. When the task or the subject is known, the system records the data by subject/task and then retrieves and ranks the unseen documents for the given subject/task.

As regards to the model described in Sect. 3.9.1, the experiments aimed to test whether the density matrices computed from a set of features observed from a sample of seen document set enable the retrieval of useful/useless unseen documents. In particular, it was of interest to test if a subset of density matrices may enable the retrieval of a subset of useful documents different from the subset of useful documents retrieved by other density matrices. If this would have been the case, different density matrices could represent different contextual factors. Moreover, when the analysis was conducted by task and/or subject, the event that the density matrices computed from the subset of data referring to a pair task/subject enable the retrieval depending on the pair could be tested. While the interest was mainly in testing whether a density matrix can enhance IRF, it was possible to know the optimal combination of features to find novel relevant information using density matrices specifically implemented in a personalized way for, say, user 1 and task 2.

3.11.4 Entanglement of User Behavior and Document Content

In Sect. 3.9.2, the quantum mechanical framework was applied to modeling the combination of the user's behavior with the content of the documents visited by the user. It was hypothesized that the user's behavior and the visited document content were entangled. In the following, a methodology for investigating this entanglement is illustrated. The methodology is in the following steps:

- Preparation of the interaction data
- Computation of a contingency matrix

- Decomposition of the contingency matrix
- Analysis of entanglement

3.11.4.1 Preparation of the Interaction Data

The log files of real subjects who interacted with a WWW browser were used to simulate a user who accesses a series of WWW pages and performs some actions such as reading, scrolling, bookmarking, and saving. The product space $A \otimes B$ can be an abstraction of a system that monitors subject behavior and utilizes these interaction data as a source of IRF to retrieve and order the unseen documents. The dataset used in this example was gathered during a longitudinal user study reported by Kelly (2004).[6] The dataset collects the data observed from seven subjects over 14 weeks and has information about the tasks performed by the subjects, the topics for which the subjects searched the collection, and the actions performed by the subject when interacting with the system. The dataset consists of a set of tuples with each referring to the access performed by a subject when visiting a webpage. The following document features of the dataset were used:

- The unique identifier of the subject who performed the access
- The unique identifier of the attempted task, as identified by the subject
- The display time, that is, the length of time that a document was displayed in the subject's active Web browser window ("display")
- A binary variable indicating whether the subject added a bookmark for the webpage to the bookmark list of the browser ("bookm")
- A binary variable indicating whether the subject saved a local, complete copy of the webpage on disk ("save")
- The frequency of access, namely, the number of times a subject expected to conduct online information-seeking activities related to the task ("accessfr")
- The number of keystrokes for scrolling a webpage ("scroll")

In addition to these features, relevance scores assigned to each document based on how relevant it was for a given task for each subject have been used. These scores were assigned by the participants in the study based on their own assessment of the relevance of the document for the task. The features were selected on the basis of their relevance to the task of IRF. In particular, "subject" and "task" allow the researchers to analyze IRF by user and by task. The other features referred to the interaction between the system and the user. In particular, "relevant" was the assessment provided by the user as regards the visited webpage, while the other features were what the system recorded about the interaction.

---

[6]I thank Dr. Diane Kelly for granting her permission to use the data.

3.11.4.2 Computation of the Contingency Matrix

The contingency matrix $\mathbf{C}$ has been computed as follows. The dataset was normalized so as to provide every feature with zero mean and standard deviation; in this way, the features were considered independently of the order of magnitude of the observed values. After this normalization, a subset of subject identifiers and a subset of task identifiers were chosen; all the subject identifiers and all the task identifiers were also used in the investigation. The five features were selected for each subject, and then the average value of every feature was computed from the tuples of the chosen subject and task identifiers. The average values were grouped by relevance score into a binary scale as follows: the documents of the tuples which recorded a relevance score between one (i.e., the minimum relevance score) and five were assessed as useless, whereas the documents of the tuples which recorded a relevance score between six and seven (i.e., the maximum relevance score) were assessed as relevant; these ranges were decided upon by the distribution of the tuples across the seven levels of relevance, about half of the tuples were distributed across the two highest scores. Finally, the 2×5 matrix was normalized so that the sum of the squares of the elements is one where 2 refers to the number of distinct relevance values and 5 refers to the number of features selected for this analysis. As an example, the matrix $\mathbf{C}$ when all the subjects and all the tasks have been selected was

$$\mathbf{C} = \begin{pmatrix} -0.162 & -0.066 & 0.020 & -0.037 & 0.004 \\ 0.201 & 0.082 & -0.025 & 0.043 & -0.005 \end{pmatrix}$$

where the column vectors correspond, respectively, to "display," "bookm," "save," "accessfr," and "scroll." The matrix $\mathbf{C}$ has then been normalized so that

$$\sum_{i,j} c_{ij}^2 = 1 .$$

The first row refers to useless documents, whereas the second row refers to the relevant documents. Of course, other groupings of the relevance scores or other features could have been used.

3.11.4.3 Decomposition of the Contingency Matrix

The contingency matrix $\mathbf{C}$ has been used for defining $|\phi\rangle$ after assuming the canonical bases $|j\rangle$ and $|k\rangle$ for A, B, respectively. The canonical basis vector $|j\rangle$ for A is the vector with 1 in the j-th coordinate and zero elsewhere; if $n_B = 2$ and $n_A = 5$, $|3\rangle$ for space A is $(0, 0, 1, 0, 0)$, where $|1\rangle \otimes |3\rangle$ of $A \otimes B$ is $(0, 0, 1, 0, 0, \ldots, 0)$. Therefore, $|i_A\rangle$ is the i-th column of matrix $\mathbf{A}$ and $|i_B\rangle$ is the i-th column of matrix $\mathbf{B}$.

After computing the SVD of the above exemplified matrix **C**, the following matrices are obtained:

$$\mathbf{A} = \begin{pmatrix} 0.902 & 0.182 & 0.108 & -0.375 & 0.023 \\ 0.367 & 0.074 & 0.033 & 0.927 & 0.007 \\ -0.110 & -0.022 & 0.994 & 0.010 & -0.001 \\ 0.198 & -0.980 & 0.000 & 0.000 & -0.000 \\ -0.023 & -0.005 & -0.001 & 0.002 & 1.000 \end{pmatrix}$$

where the column vectors, respectively, correspond to $|1_A\rangle, |2_A\rangle, |3_A\rangle, |4_A\rangle, |5_A\rangle$,

$$\mathbf{B} = \begin{pmatrix} -0.629 & 0.777 \\ 0.777 & 0.629 \end{pmatrix}$$

where the column vectors, respectively, correspond to $|1_B\rangle, |2_B\rangle$, and

$$\Lambda = \begin{pmatrix} 0.9999 & 0.0000 & 0.0000 & 0.0000 & 0.0000 \\ 0.0000 & 0.0059 & 0.0000 & 0.0000 & 0.0000 \end{pmatrix}$$

Thanks to orthonormality provided by the SVD, the columns of **A** and **B** are a basis of space A and B, respectively. The columns of the three matrices are ordered by decreasing "importance" where the importance of the columns is measured by the corresponding singular value in the diagonal of Λ. The importance is given by the fact that the top k singular values and the corresponding columns of **A** and **B** provide an approximation of **C**; this notion is similar to the one exploited in the spectral theorem and was previously exploited in IR when LSA was proposed by Deerwester et al. (1990) for computing a reduced representation of a document collection. Indeed,

$$|\phi^{(k)}\rangle = \sum_{i=1}^{k} \lambda_{i,i} |i_A\rangle \otimes |i_B\rangle$$

is as close to $|\phi\rangle$ as k is close to the number of nonzero singular values. As the columns of the three matrices are ordered by the singular values, the contribution of $|i_A\rangle \otimes |i_B\rangle$ decreases as k increases, that is, the first columns of **A** and **B** gives the greatest contribution to the approximation of **C**.

The pattern of the values in every column vector provides an interpretation of the role played by the vector in the context of the implicit feedback framework illustrated above. The column vectors of **A** explain the various ways in which the documents were visited by the subjects; let us name it "behavioral factor." In particular, $|1_A\rangle$ states that "display" was the most influential feature of the first behavioral factor, while "bookm" was the second most influential. Indeed, the weight of "display" is close to 1. If $|1_A\rangle$ was a state of a system, and the canonical basis vector $|1\rangle$ of A was used to compute a projector, then the square of the first coordinate of $|1_A\rangle$ would be the probability that "display" was determining user behavior.

Similarly, the column vectors of **B** explain the various ways in which the subjects assessed the relevance of the visited documents; let us name it "relevance factor." In particular, $|1_B\rangle$ states that the relevance scores were slightly predominant since $|0.777| > |-0.629|$ in the first relevance factor. The other information carried by this relevance factor is that the tendency of assessing documents as relevant was negatively correlated with the tendency of assessing documents as useless. This information should really not be surprising; however, the examples illustrated in the following show that this correlation is not always the case.

3.11.4.4 Analysis of the Entanglement

Using the Schmidt number (Sect. A.8), the previous example shows $\lambda_{1,1} \approx 1$. As a consequence,

$$|\phi\rangle \approx |1_A\rangle \otimes |1_B\rangle$$

thus showing that entanglement is virtually absent in the example. From an implicit feedback point of view, the example states that when all the subjects and the tasks are considered in computing the contingency matrix, the resulting state vector of $A \otimes B$ is a product state vector, i.e., it is separable into $|1_A\rangle$ and $|1_B\rangle$. This means that the way in which the documents are visited is in practice independent of the way in which the documents are assessed, when the systems are observed without reference to the subject and to the task actually performed.

The hypothesis that may in general be stated is that entanglement is somehow related to the subject or to the task, that is, the way a document is visited by this subject who is performing this task is entangled with the way the subject assesses the relevance of the document. With the aim of checking whether entanglement is related to the subject or to the task, the contingency matrix only for subject 1 and task 1 has been computed:

$$\mathbf{C} = \begin{pmatrix} -0.158 & -0.094 & -0.052 & -0.110 & 0.835 \\ 0.014 & 0.128 & -0.052 & 0.256 & 0.411 \end{pmatrix}$$

After decomposition, the following three matrices for subject 1/task 1 are provided:

$$\mathbf{A} = \begin{pmatrix} 0.145 & 0.243 & -0.053 & -0.188 & 0.939 \\ 0.031 & 0.473 & 0.092 & -0.827 & -0.288 \\ 0.074 & -0.074 & 0.991 & 0.045 & 0.073 \\ -0.011 & 0.843 & 0.048 & 0.525 & -0.109 \\ -0.986 & 0.035 & 0.069 & -0.057 & 0.135 \end{pmatrix}$$

$$\mathbf{B} = \begin{pmatrix} 0.90 & -0.43 \\ 0.43 & 0.90 \end{pmatrix}$$

$$\Lambda = \begin{pmatrix} 0.94 & 0.00 & 0.00 & 0.00 & 0.00 \\ 0.00 & 0.33 & 0.00 & 0.00 & 0.00 \end{pmatrix}$$

These three matrices are rather different from the respective matrices computed when all the subjects and all the tasks have been selected. First, the most important column states that the behavior of subject 1 when visiting the documents and performing task 1 is very different from the average behavior since scrolling appears to be the most important feature. Moreover, this subject does not tend to assess the documents either as relevant or irrelevant; in fact, he has a more "balanced" assessment than the average subject. The most interesting outcome of this example is that the Schmidt number is significantly greater than 1 because the second singular value is significantly greater than 0. This outcome implies that $|\phi\rangle$ is entangled, that is, cannot be separated into one state about the document and one about the user.

In general, there are many subject/task pairs for which the computed contingency matrix yields an entangled state $|\phi\rangle$. This outcome suggests that in an implicit feedback environment, the behavior of the user when assessing the relevance of a document is entangled with the way the document is visited. This outcome has been confirmed in previous, independent experiments, thus suggesting that the context in which the interaction takes place should be considered when the information retrieval system has to rank and present the documents to the end user.

3.11.5 Concept Combination

Aerts and Gabora (2004a) reported an emblematic study of the quantum mechanical framework in the field of Natural Language Processing (NLP), the latter being a topic relevant to and overlapping with IR; another empirical study was reported by Bruza et al. (2012).

The study reported by Aerts and Gabora (2004a) was performed with the collaboration of 81 human subjects who were asked to answer a written questionnaire. In this questionnaire, the notions underlying SCOP (Sect. 3.4.2) were utilized. The subjects were given a list of exemplars. For each exemplar and for each context, a subject provided a subjective frequency with which the exemplar would fit the context. After collecting the assessments by the subjects, it was possible to estimate the probability that an exemplar of a concept is assigned to a context; for example, the probability that an exemplar (e.g., "dog") of *pet* is assigned to the ground context could be estimated (e.g., the probability that "dog" is a exemplar of *pet* was 0.12). Varying the context, different probabilities of the same exemplar could be estimated; for example, the probability that "dog is chewing a bone" was much higher than "canary is chewing a bone" and higher than "dog is being taught" since overall the subjects thought that a dog chews a bone more frequently than a canary and more frequently than being taught. Moreover, varying the context, different probability distributions could be obtained—each distribution was conditioned by a context—without any relationship with the conditional probability distribution obtained when the ground context is considered, that is, the probability that an exemplar is in the ground context is not the sum of the probabilities that the exemplar is in the individual contexts. Therefore, the ground context is the linguistic union of the individual contexts, but this union is neither algebraic nor mathematical; given that $e_1, \ldots, e_6$ were six contexts such that $P(e_1) + \cdots + P(e_6) = 1$ and x was an exemplar,

$$P(x \in e_1)P(e_1) + \cdots + P(x \in e_6)P(e_6) \neq P(x \in \text{ground context})$$

where $x \in e$ means that x is pertinent to context e, but the ground context cannot be partitioned in $e_1, \ldots, e_6$.

To observe the "guppy effect" described in Sect. 3.4, the experimental study had to provide the necessary data. To this end, Aerts and Czachor carried on the experiment using the same subjects and the data acquisition methods based on concepts, exemplars, and contexts. In particular, the subjects were asked to assess the pertinence of the exemplars provided in a list to the context "the *fish* is a pet" where *fish* plays the role of concept and "pet" contextualizes the concept; for example, it was found that "guppy" was an exemplar pertinent to the context (i.e., "guppy is a pet") with relative frequency 0.40 and that "goldfish is a pet" with relative frequency 0.39. Note that the frequencies of membership to the context "the *fish* is a pet" would be different from the frequencies of membership to the context "the *pet* is a fish" where the roles of "pet" and "fish" are swapped. Moreover, other contexts in which *pet* and *fish* appear to be entangled were defined and submitted to the subjects; for example, "The pet swims around the little pool where the fish is being fed by the girl" was such a context. It is in this context that the probability that "guppy" was pertinent to the context was higher than the probability that it was pertinent to the contexts involving *pet* or *fish*, thus suggesting that concepts can be combined in a way that the quantum mechanical framework and in particular entanglement can model this combination.

3.11.6   Semantic Spaces

Bruza and Cole (2005) used semantic spaces to experimentally investigate collapse. A particle is usually not in a pure state of whatever observable (e.g., momentum) is intended to be measured. When the observable is measured, the state of the system will immediately become a pure state of that observable. This process is known as collapse (see also Sect. 2.2). The empirical investigation reported in that paper may be reproduced according to the guidelines utilized when LSA was experimented in the 1990s since collapse resembles what happens when eigenvectors are used to retrieve documents.

Collapse can be investigated in semantic spaces because when a word is viewed in its context, it is represented by a mixed state and a measurement of the word's sense is made on this word; the mixed state of the word collapses onto one of its senses which is given by the word's context. The senses of a word are the observables. For example, when word w is seen in the context of sense t, the state vector $|w\rangle$ collapses onto the pure state representing the sense dealing with t. After collapse, weights of associations to words related to w will be high, whereas before collapse, the weights of such associations may have been weak. The associations may become more intense and approach the point of being expressed. The description above of the relationships between context and collapse is essentially the same as that of Aerts and Gabora (2004b).

Suppose a word w is given and a HAL matrix can be computed with respect to a word as illustrated in Sect. 3.6. This matrix can be computed considering all and only the words around the word that occur in a long document or in a collection of documents; the words around the word would represent the various contexts of the word. This matrix is therefore a numerical summary of the contexts of the word in the collection of documents. In the experiments, the word was "Reagan" and the document collection was Reuters-21578. The corpus was scanned by text windows of fixed length which are centered around "Reagan." For each window, it is possible to construct a matrix similar to the matrix illustrated in Sect. 3.6; each entry reports the distance between a word of the window and "Reagan." It is crucial that the dimensions of the matrix are fixed to a value which is the number of words considered; these words may be included in a predefined vocabulary, for example, the vocabulary built from the ranked list of documents retrieved against the one-word query "Reagan" depicted in Fig. 3.23. All these matrices are summed to obtain a matrix of the same dimensionality that represent the semantic space of "Reagan." The matrix embeds the different senses of the word "Reagan."

To extract the different senses of the word "Reagan" from the matrix, SVD is utilized. The eigenvectors are a representation of these senses, and the eigenvalues measure the significance of each sense. To obtain sensible eigenvectors, a great deal of attention should be paid to the selection of the vocabulary words; for example, those displayed along with "Reagan" in Fig. 3.23 might not be very meaningful as those used by Bruza and Cole (2005) who reported the eigenvectors of Fig. 3.24. The values associated to a word of an eigenvector measure the closeness between "Reagan" and the word. The first eigenvector is a sort of summary of the words occurring with "Reagan." The other three eigenvectors represent alternative (i.e., orthogonal) senses; for example, the second eigenvector represents the sense of government, while the third eigenvector represents the sense of export. The interpretation is left to the humans using these eigenvectors.

Consider collapse. Suppose the word "Iran" can be described as a state vector; the actual implementation of this state vector can begin from the semantic space of "Iran" built using the same procedure used for "Reagan." The first eigenvector of the matrix that represents the semantic space of "Iran" can provide the state vector. On

Reagan said. Reagan said the message that the nation should move on had come from
Reagan's veto of the road bill. Schneider contrasted Reagan's failure with his previous
Reagan was ignorant about much of the Iran arms deal, just about ends his prospects
Reagan officials said the president is ready to fight them all. Reagan said in announcing
Reagan celebrate their 35th wedding anniversary today as Reagan prepares to make an address to the nation
Reagan would sign into law. Reagan last year blocked Senate consideration of a tough House
Reagan said. Asked whether he would consider raising taxes, Reagan said I am willing to look
Reagan would sign into law. Reagan last year had blocked Senate consideration of a tough
Reagan nominated economist Alan Greenspan in his place. Volcker's term expires in August. Reagan
Reagan and his new chief of staff today begin trying to revive an administration tattered
Reagan's denial that he knew proceeds from Iran arms sales were diverted to Nicaraguan
Reagan, 76, to demonstrate he was in command of affairs despite the crisis. Reagan said
Reagan said he would discuss the Mideast Gulf situation with allied leaders at next week
Reagan administration that he stay at the helm of the U.S. central bank, U.S. officials
Reagan expected to focus on superpower arms control moves and trade issues. French officials said
Reagan is vacationing. In his March annoucement, Reagan said I am committed to full enforcement
Reagan's budget was clobbered in an early vote in the House. When Reagan entered
Reagan said in a statement following a meeting with his top economic advisers. Reagan said
Reagan, fighting to recover politically from the Iran-contra scandal, plans to acknowledge in a critical
Reagan's speech on the Iran arms scandal as candid and constructive while Democrats, who control
Reagan to join in a bipartisan effort - or even a summit - to write a new budget
Reagan on two occasions in 1986 that profits from arms sales to Iran were being
Reagan's statement followed a meeting with his top economic advisers. Reagan said he remains
Reagan said he hoped the United States could lift trade sanctions against Japan soon. But he said
Reagan. Last Friday Reagan, answering questions following a speech to the Los Angeles World Affairs
Reagan is concerned about the record drop in stock prices but remains convinced on the basis
Reagan warned the U.S. Congress in his weekly radio address against passing what he called
Reagan whose autocratic rule in the White House angered some top Reagan officials and, perhaps
Reagan said. Reagan added: Rest assured, there's plenty of record-keeping now going on at 1600 Pennsylvania
Reagan said the United States would offer a draft treaty on medium-range missile reductions

Fig. 3.23 Reuters-21758 documents about Reagan

reagan (0.62), president (0.48), administration (0.22), house (0.17), trade (0.15), congress (0.11), budget (0.11), bill (0.10), veto (0.10), white (0.09), tax (0.09), japan (0.08), senate (0.08), billion (0.08), iran (0.07)
reagan (0.74), ... bill (-0.04), congress (-0.05), trade (-0.07), house (-0.08), administration (-0.23), president (-0.55)
japan (0.25), trade (0.25), japanese (0.24), tariffs (0.21), administra- tion (0.13), united (0.11), sanctions (0.11), exports (0.11), ... tax (-0.11), senate (-0.13), veto (-0.14), budget (-0.19), white (-0.31), house (-0.38)
billion (0.44), dlrs (0.37), dlr (0.21), budget (0.18), veto (0.18), deficit (0.17), bill (0.14), highway (0.13), mln (0.10),...conference (-0.07), house (-0.08), baker (-0.09), scandal (-0.12), white (-0.14), arms (-0.24), iran (-0.25)

Fig. 3.24 Senses of "Reagan"

the other hand, the eigenvectors of the matrix that represents the semantic space of "Reagan" can correspond to the outcome of the observable used to measure sense; whenever sense is measured, the outcome is one of the eigenvectors of the matrix that represents the semantic space of "Reagan." Therefore, the probability that the sense "Reagan" can be observed is given by the trace rule applied to the eigenvector of the matrix that represents the semantic space of "Reagan" and the state vector of "Iran." The former will be the result of the collapse of the latter. The probability that is given by the trace rule is then the probability of collapse on the observed sense.

3.11.7 Quantum Probability Ranking Principle

The quantum PRP has been introduced in Sect. 3.8, formerly proposed by Zuccon et al. (2009) who performed an evaluation in 2010. The evaluation of the quantum PRP requires the estimation of the interference term. Zuccon and Azzopardi (2010) estimated the interference for the task of subtopic retrieval with the aim of capturing the document interdependence through the interference term. The basic idea of those authors is to encode document interdependence using the cosine of the angle that appears in the interference term, whereas the real part acts as an additional weight and is estimated by the probabilities of relevance. As these probabilities are multiplied when the interference term is calculated, the latter is a monotonically increasing function of the two probabilities.

However, the interference also depends from the cosine of the angle of a complex number. As the estimation of this angle is quite complex because it is still unclear what it can represent, the authors estimated it to measure the interaction between documents covering different facets of the topic in the subtopic retrieval task; in particular, the authors assumed that redundant relevant documents negatively interfere, while documents conveying relevant but novel information generate positive interference. They implemented the quantum PRP by means of an approximation of the cosine of the angle by using the Pearson correlation index computed between the document vectors. The experimental results, which are reported in the paper, were encouraging.

3.12 Suggested Readings

van Rijsbergen (2004) proposed to utilize the quantum mechanical framework for integrating the logical, the probabilistic, and the vectorial approaches to IR. Although the mathematical framework utilized is the same as the framework of QM, the main aim of that book is to show how the logical, probabilistic, and vector space views can be combined in one mathematical framework. In particular, he proposed kinds for an alternative, nonclassical logic and pointed out the Gleason theorem to explain that the logical, probabilistic, and vector space views can be combined in one mathematical framework.

Khrennikov (2010) addressed the problem of mapping a probability space to a state vector and proposed the Quantum-Like Representation Algorithm (QLRA) to this end. Suppose two binary observables A, B, the conditional probabilities $P(A = a|B = b), P(B = b|A = a)$ and the marginal probabilities $P(A = a), P(B = b)$ are given. The problem is to find one state vector ϕ and the projectors $|a\rangle\langle a|, |b\rangle\langle b|$ over the complex field such that

$$P(A = a) = |\langle \phi|a\rangle|^2 \qquad P(B = b) = |\langle \phi|b\rangle|^2$$

The QLRA provides a solution to this problem.

Kinds were introduced by Hardegree (1982). To this end, he defined natural classes as follows: a class is a natural class if and only if there are traits that all, and only, the class elements share. After this, Hardegree made a connection between the functions tr(.) and in(.) and a Galois connection and then to logic. Finally, he introduced the logic of kinds. Kinds might be implemented by the itemset mining algorithms proposed by Agrawal et al. (1996) and Han et al. (2000). The implementation of kinds can be based on techniques other than itemset mining; for example, bi-clustering and co-clustering are relevant.

The literature on concept combination is mainly formed by the publications by Aerts and Gabora (2004a,b), Aerts (2009), Bruza et al. (2010), Bruza et al. (2012), and Busemeyer and Bruza (2012). Aerts and Gabora (2004a) proposed the concept theory called state-context-property theory (SCOP), which is an application of the quantum mechanical framework to the context and property of natural language; for example, the notion of state of a concept is introduced in this paper. Aerts and Gabora (2004b) explained how to use SCOP to model context and to explain how context influences the state of a concept by means of the pet-fish problem. Bruza et al. (2010) investigated the hypothesis that word associations can display "spooky action at a distance behavior." They suggested to use this action, which was investigated in the quantum mechanical framework, for modeling the human mental lexicon and for comparing spreading activation with the results obtained from experiments on associations between words to be recalled before reading other words. Bruza et al. (2012) investigated how human subjects produce associate words (e.g., "slave") in relation to combinations (e.g., "pet human") although these associate words cannot be produced as associate words of the words (e.g., "pet" or "human") in isolation. They argued that these associations are formed below the symbolic level of cognition, and to this end, they proposed to utilize the quantum mechanical framework; in particular, the authors showed and empirically tested how concept combinations may behave like entanglement. Busemeyer and Bruza (2012) highlighted how the understanding of human thinking is based on probabilistic models and argued that the underlying quantum mathematical framework provides an account of human thinking. They introduced "contextuality," as a way to understand interference effects found with inferences and decisions under conditions of uncertainty. They also proposed to use entanglement to model cognitive phenomena.

The use of the quantum mechanical framework to model word ambiguity and to introduce vector negation was proposed in a series of papers authored by Widdows and Peters (2003), Widdows (2004), Widdows and Cohen (2010). Widdows and Peters (2003) combined the geometric structure of vector models with Boolean logic for modeling natural language. The authors formulated logical connectives in vector spaces based on standard linear algebra; in particular, they explained vector negation to disambiguate words. It is this work that replaced sets by subspaces of a vector space and set unions, intersections, and complements by vector sum, intersection, and orthogonal complements of subspaces. The book by Widdows (2004) gathers the results of the previous papers and some chapters on logic, NLP, and QM together. Widdows and Cohen (2010) illustrated the application of the semantic vector package to understanding the informative content of a textual document collection from the distribution and usage of words in a corpus of text.

Semantic spaces were proposed by Bruza and Cole (2005). Bruza and Cole highlighted the scarcity of large-scale experiments and scalable computer systems for intelligent NLP. Starting from Gärdenfors (2000), they leveraged his idea that at the conceptual level, "properties and concepts have a geometric representation in a dimensional space. For example, the property of 'redness' is represented as a convex region in a tri-dimensional space determined by the dimensions hue, chromaticity and brightness." They conjectured that the theory of vector spaces can implement the low level below the conceptual level at which concept formation and combination take place. It is within this context that HAL was utilized as described in Sect. 3.6.

A principled way to address contextual IR has been proposed by Melucci (2008a). That paper utilizes the quantum mechanical framework to present a principled approach to modeling context and its role in ranking information objects using vector spaces. First, he outlined how a basis of a vector space naturally represents context, both its properties and factors. Second, a ranking function computes the probability of context of the objects represented in a vector space, namely, the probability that a contextual factor has affected the preparation of an object.

The quantum version of the PRP was illustrated by Zuccon et al. (2009) and further developed by Zuccon and Azzopardi (2010). Zuccon et al. (2009) reformulated the PRP based on the quantum mechanical framework, which explicitly models interference between observables, in particular, between relevance assessments. The outcome is the quantum PRP. Later, Zuccon and Azzopardi (2010) explored whether the quantum PRP leads to improved performance for subtopic retrieval, where novelty and diversity are required.

The use of vector spaces for modeling user interaction was proposed by Melucci and White (2007), while Piwowarski et al. (2010) proposed these spaces for modeling information needs, and Frommholz et al. (2010) proposed them for modeling polyrepresentation. Melucci and White (2007) presented a geometric framework that utilizes multiple sources of evidence present in this interaction context (e.g., display time, document retention) to develop enhanced implicit feedback models personalized for each user and tailored for each search task. They used rich interaction logs (and associated metadata such as relevance judgments), gathered during a longitudinal user study, as relevance stimuli to compare an implicit feedback algorithm developed using the framework with alternative algorithms. Piwowarski et al. (2010) presented an approach based on the quantum mechanical framework to addressing different issues of IR. They described some of the properties of this framework when representing queries and documents. Then, they investigated and experimented tensor products with different parameters using Text REtrieval Conference (TREC) collections. Frommholz et al. (2010) discussed how the quantum mechanical framework can be extended to support polyrepresentation. They also illustrated how polyrepresentation can be implemented by means of tensor products.

The use of quantum detection theory for finding optimal subspaces and defining optimal ranking was proposed by Melucci (2012a). As indexing cannot be exhaustive nor precise, the decision taken by an IR system about the relevance of the content of a document to an information need is subject to uncertainty. He stated the hypothesis that the system is unable to optimally respond to every query because the document collections and the posting lists are modeled as sets of documents and proposed to replace sets by vector spaces. Using the connection between Neyman-Pearson Lemma (NPL) and PRP, he defined a Quantum Information Retrieval Basis (QIRB) that at least in principle yields more effective document ranking than the ranking yielded by the current principles, with effectiveness being measured in terms of recall at every level of fallout.

An exhaustive and detailed introduction to evaluation of interactive IR systems utilized by real users was written by Kelly (2009). She provided a historical background on the development of user-centered approaches to and the major components of the evaluation of interactive IR systems. She also described different experimental designs and sampling strategies, presented core instruments and data collection techniques and measures, explained basic data analysis techniques, and discussed validity and reliability issues with respect to both measures and methods.

Chapter 4
Future Work

This book aims to illustrate how the quantum mechanical framework has been applied to IR. To this end, we placed the book in the intersection between IR and QM. Then, we tried to draw a comparison between the way QM is used to bridge the gap between the values observed by means of a device and the reality of subatomic particles on the one hand and the way in which it might be used in IR to bridge the gap between relevance and content or between information need and request on the other. In this concluding chapter, some research directions are briefly outlined, thus hoping that the quantum mechanical framework will be fully leveraged to achieve effective and efficient IR systems.

4.1 Incompatibility Between Observables

Things are incompatible when they are so different in nature as to be incapable of coexisting. In QM, incompatibility is applied to observables, and thus two observables are incompatible when they are so strong in measurement or the object under measurement is so weak as to be incapable of emitting values without disturbing each other. Examples are momentum and position measured in a particle; when the particle is macroscopic (e.g., Earth), incompatibility cannot be observed and the measurement of momentum cannot interfere with the measurement of position; in contrast, incompatibility can be observed when the particle is microscopic.

In IR, incompatibility is less obvious since the particles are macroscopic; for example, a user or a document are macroscopic objects and the measurement of an observable about a physical property of the user or of the document cannot interfere with another observable. However, incompatibility might be observed at the level of user cognition and in particular of the user's information need. In principle, the user's information need can be represented by a state vector or in general by a density matrix expressing the probability distribution of the observables

© Springer-Verlag Berlin Heidelberg 2015
M. Melucci, *Introduction to Information Retrieval and Quantum Mechanics*,
The Information Retrieval Series 35, DOI 10.1007/978-3-662-48313-8_4

applied to the need. The user's information need might be considered a fragile entity which can be changed by an observable to an extent that another observable can be disturbed. Examples are relevance, aboutness, authority, quality, or other properties of a document examined by the user; if two observables, say relevance and quality of a document, were found incompatible, it would be impossible to measure one observable without measuring the other observable with intrinsic imprecision, and the measurement of one observable would not commute with the measurement of the other observable.

The discovery of incompatibility in IR would require the utilization of the logic represented by abstract vector spaces instead of the logic represented by subsets. Indeed, the logic represented by abstract vector spaces is not commutative, and the distributive law does not hold. Such a logic would therefore be necessary to model the commutativity of the observables applied to the level of the user's cognition. If incompatibility were proved and the abstract vector spaces were necessary, the use of probability would differ from the use of probability in the classical IR since the latter relies on the Kolmogorovian theory of probability described by Kolmogorov (1956). It follows that the traditional ranking principles should be replaced by principles based on non-Kolmogorovian theories such as those requiring the interference term. A new retrieval model would then be formulated.

The discovery of incompatibility in IR would require careful experimentation. In particular, the involvement of human subjects as users should be necessary. In this situation, the control of the experiments has to guarantee that the values observed from the user's interaction are effectively an expression of the occurrence of incompatibility. How this control can be achieved was explained by Ingwersen and Järvelin (2005) and Kelly (2009). When the experiments operate at the level of the user's cognition (e.g., the user's information need), some attention should be paid to the psychological issues of user interaction.

4.2   Entanglement of States

Entanglement refers to the situation in which two things are twisted together. In QM, the "things" are particles (e.g., photons) or more precisely the "things" are the states of the particles. If entangled, the particles react in the same way when measured by the same observable whatever the observable is; for example, two entangled photons reveal the same polarization although they are placed at a very long distance from each other, and the polarization of one photon changes at random.

In IR, entanglement is less apparent than in QM since it can be detected only if a single element (actually, one pair of entangled elements) is measured. In contrast, correlation can be measured when an ensemble of elements is measured as explained in Sect. 3.2.7. In IR, it is customary to collect documents, index terms, users, and other objects in ensembles such as document collections, index term vocabularies, and user communities. When these ensembles are built, some statistics (e.g., sums, means, or other distributional properties) can be computed. If one single element of an ensemble is available, these statistics are not justifiable. As correlation requires means and covariances, correlation cannot be computed for one single element unless one accepts trivial correlations; for example, the measurement of height and weight of one person will always give perfectly correlated measures.

The discovery of entanglement in some IR processes or objects would open up into a situation in which the measurement of correlation can be feasible even within one or a few elements; for example, the measurement of correlation between two observables might be made on only one user at different instants of time or between two users who have been in contact at one instant of time and then have been placed at two far places and observed at future instants. To test whether a pair of elements is entangled, the statistical test based on Bell's inequalities can be utilized. The problem is that entanglement cannot be measured on the basis of a number of identical elements subjected to the same observables; for example, a pair of users has to be put in an entangled state, but how this preparation has to be done is an open problem.

A connection between this situation and Bayesian statistics may be established. It is accepted that probability is likely to cause an argument since two views are confronted with the interpretation of the mental experiments envisaged when the probability space is defined. On the one hand, according to the frequentist view, probability can only be estimated by repetitions of an experiment; for example, in coin tossing, the probability of heads might be obtained as the experiment of flipping a coin can be repeated so as to calculate the limit of the number of heads that occurred over the number of tosses. On the other hand, in some cases, potential repetitions of an experiment cannot be envisaged so that the long run of probability cannot be defined with respect to a potentially infinite repetition of experiments; for example, based on the assessments made by two users about the relevance of a few documents, an IR system tries to estimate the probability that the users will think relevant on each of the documents in the collection. The definition of the correlation between the users as a limiting case of infinite repetitions of the same experiment would not make much sense since we cannot repeat the experiment. However, if we assume that the users behave in a manner consistent with other pairs of users, we should be able to exploit the large amount of data from other pairs of users to make a reasonable guess as to what correlation exists. This degree of belief or Bayesian subjective interpretation of probability avoids non-repeatability issues.

4.3 Relevance Detection

In Sect. 3.10, we explained that the identification of the observables corresponding to the eigenvectors is a hard task. When vector subspaces are utilized for implementing compatible observables, the projectors must commute. However, in principle, the definition of projectors that cannot commute is possible, and indeed the optimal eigenvectors defined in Sect. 3.10 correspond to noncommutative projectors. As a matter of fact, the physical interpretation of these noncommutative projectors is rather difficult—they are a kind of immaterial feature—yet an axiomatic definition of the optimal projectors is still possible; for example, the optimal projectors must meet some necessary conditions, for instance, the corresponding events must be such that the distributive law does not hold when they are combined with the projectors $|0\rangle\langle0|$ and $|1\rangle\langle1|$ of the classical, physical observables such as index term occurrence. Moreover, the optimal projectors correspond to subspaces oblique to the subspaces spanned by $|0\rangle, |1\rangle$ at given angles. However, some practical techniques are necessary other than axiomatic definitions.

One direction is to explore the potential of LSA illustrated in Sect. 1.3 and other applications of matrix decomposition techniques to check whether the semantic vectors extracted from a text can implement the eigenvectors; the semantic vectors have been studied within the research surveyed in Sects. 3.5 and 3.6. The subspaces spanned by the projectors obtained by matrix decompositions are usually oblique to the subspaces spanned by the basis vectors used to represent the classical, physical observables; besides LSA and SVD, it is worth mentioning Non-negative Matrix Factorization, Principal Component Analysis, and Factor Analysis (see, for example, the book by Bartholomew et al. (2008)).

It would also be possible to exploit the kinds introduced in Sect. 3.3 to illustrate how indexing and retrieval can be axiomatized and defined in terms of kinds instead of the classical content descriptors managed by an IR system. Kinds may exhibit characteristics useful for meeting the axiomatic requirements to identify the observables corresponding to the optimal projectors. Indeed, we have shown that there exist three kinds that cannot meet the distributive law of meet and join between kinds. The problem is to find an algorithm for mapping projectors with kinds, obtaining a representation of the optimal projectors as kinds, and defining projectors applied to subspaces as the application of meet and join to kinds.

4.4   Computing and Ranking Kinds

In Sect. 3.3, kinds have been introduced to provide a logic of representation of the informative content of documents alternative to classical content descriptors (e.g., index terms), which provide a "linear" description of content in the sense that the list of documents indexed by a descriptor is a "line" of document identifiers and statistics of the occurrence of the descriptor in the documents. Kinds provide a "rectangular" description of content in the sense that documents (i.e., individuals) and index terms (i.e., traits) form a rectangle of pairs of documents and index terms.

From a practical point of view, the collection of kinds ordered by probability of relevance should be presented to the end user according to the same principle utilized to rank documents by probability of relevance. However, the user is used to receiving ordered lists of documents and is not used to receiving kinds; for example, the current search engines deliver ordered lists of WWW pages. Presenting the kinds to the end user by presenting the individuals (i.e., the documents) in order of the $P_1(K)$s would be a convenient option. This stratagem would allow the researcher to employ the widely adopted methodologies of laboratory-based experimental studies such as those based on test collections, which require lists of documents in order of measures (e.g., probability) of relevance. However, such a stratagem nevertheless ignores the presentational issues of the ordered collections of kinds which are likely to require innovative approaches to the general problems encountered within the research in information access and seeking. The problem of ranking and presenting kinds to the end user is thus connected to the problem of ranking and presenting clusters or other conglomerates of data which can be described in terms of graphs.

Another development of the theory of kinds described in Sect. 3.3 would aim to implement the probability of kind (3.4). The formulation of the probability of kind, that is,

$$P(K) = P(A)^{s(T)} \qquad K = (A, T)$$

connects with the BM25 weighting scheme. Suppose we are given two hypothesis— relevance and nonrelevance to an information need—and we would like to test the hypothesis that a kind contains information relevant to the information need against the hypothesis that the kind does not. Let $P_1(A)$ be the probability that A is observed under the hypothesis of relevance and $P_0(A)$ be the probability that A is observed under the hypothesis of nonrelevance. The log-likelihood will be

$$\log \frac{P_1(K)}{P_0(K)} = s(T) \log \frac{P_1(A)}{P_0(A)}$$

where $s(T)$ connects with the saturation component and the logarithm connects with the TRW component of BM25. How to estimate $s(T)$ and the likelihoods is a matter of future research.

4.5 Alternative Logics for User Interaction

Usually, users of an IR system submit textual queries implemented as bag of words, that is, lists of words without Boolean operators; the retrieval system may automatically add conjunction and disjunction operators to control the quantity of retrieved documents. The utilization of Boolean operators implies the classical set-based approach to representing information; words are subsets of documents and the operators are set operations as explained in Sect. 1.2. This approach might reveal many limitations when users are experts who perform complex search tasks; for example, a journalist may want to learn something more about the debate on the use of nuclear power; after searching the WWW by using some standard search engine, she may realize that the theme (e.g., topic, concept, subject matter, motif, argument, thesis) is much more complex than she has expected and strongly dependent on time, application, and country—in such a situation, searching using themes would be more effective than finding WWW pages using words.

Searching using themes may require a logic different from the classical logic applied when finding documents using words. Indeed, a theme can be made of words; however, these words may not be associated by set membership but by subspace membership, thus introducing the need of processing information using subspaces. In the previous chapters, it was explained that processing information using subspaces implies a logic different from the classical logic; for example, the distributive law does not hold. Therefore, novel operators that work with subspaces are needed.

These operators would aim to provide the end user with a language for expressing themes, combining themes through operators and evaluating the degree to which a document is relevant to a theme. In particular, they would aim to allow the end user to express the fact that a theme can be joined with another theme and that the result is spanned by both themes. Moreover, these operators would aim to allow the end user to express the fact that two themes are met together and that the result spans both themes. The end user would then be provided with join and meet operators such that, if t_1 and t_2 indicate themes, the join of two themes can be expressed by $t_1 \wedge t_2$ and the meet of two themes can be expressed by $t_1 \vee t_2$.

If an artificial language has to be defined, the following artificial commands of this language may give an adequate correspondence so that the join and the meet of "nuclear power" and "green economy" can be written as "nuclear power JOIN green economy" and "nuclear power MEET green economy." Informally, the meet is the "smallest" theme spanning both themes, and the join is the "largest" theme spanned by both themes. For example, "java JOIN computing" would be the "largest" theme spanned by both themes, and, if a name is needed, "java programming language" may work since it spans "java," which is an aspect of the Java language, and "programming language," which is more general than the Java language; as another example, "abstract data type MEET inheritance" would be the "smallest" theme spanning both themes, and "object-oriented programming" may be a name of this theme since "abstract data type" and "inheritance" are aspects of the object-oriented paradigm. The analogy between join and meet, from the one hand, and intersection and union, on the other hand, may be promptly noted—the crucial difference is that when themes are implemented by constructs other than the sets used to implement union and intersection, the classical distributive law fails as shown in Sect. 3.2.

4.6   Multimedia and Multimodality

In Sect. 3.3, we mentioned that the use of index terms extracted from texts implies the utilization of intersection or union of the posting lists, thus implementing classical retrieval. The naturalness of this set-based approach to text retrieval is ascribable to the easy recognition of index terms in documents and to the user's intuition that an index term corresponds to a set of documents. When terms are combined by Boolean operators, it is assumed that an index term has a semantics.

The use of a set-based approach to retrieval with non-textual documents is far more complex. When image, video, or sound documents are to be indexed, the content descriptors are not conveniently available as index terms, and the assumption that the intersection or union of posting lists can express aboutness does not seem as intuitive as it is for text. Similarly, when click-through and user interaction data are collected, sets are not the most obvious representation of informative content. The reason is that the language of non-textual traits is likely to describe individuals with a logic which will be different from a classical logic. The use of a set-based approach to retrieving and ranking may be inadequate with multimedia or multimodal data, since the same reasons that make the retrieval of non-textual documents more complex than the retrieval of texts still hold when the criteria of acceptance of the hypothesis of relevance have to be defined. The complexity of dealing with non-textual documents, on the one hand, and the different language provided by, for example, the kinds, on the other hand, suggest that a set-based approach to retrieval should be abandoned when multimedia and multimodal data are considered.

Appendix A
Other Topics of Quantum Mechanics

A.1 Representing Qubits

Although complex vectors are a common way to represent qubits, there are alternative ways to represent qubits defined in the bidimensional space. These different representations can be more or less convenient ways to represent qubits according to the context in which qubits are used. Nevertheless, it is crucial to understand what the differences between these representations are and when they are equivalent. To this end, it is necessary to understand something more about the state space in which qubits are represented.

When we consider photons, the state space of the qubit corresponding to the photon is the set of the possible polarization values. If a photon has two polarization values in its state space, the state space consists of the possible linear combinations of the polarization values of the photon. As the polarization values correspond to the vectors $|0\rangle$ and $|1\rangle$, the state space is given by the set

$$\{\alpha_0|0\rangle + \alpha_1|1\rangle\}$$

where α_0, α_1 are two complex scalars such that $|\alpha_0|^2 + |\alpha_1|^2 = 1$. For each state of a state space of a qubit, it is possible to make a correspondence between a vector and a complex number, that is,

$$\alpha_0|0\rangle + \alpha_1|1\rangle \quad \text{corresponds to} \quad c = \frac{\alpha_1}{\alpha_0}$$

and

$$c \in \mathbb{C} \quad \text{corresponds to} \quad \frac{1}{\sqrt{1 + |c|^2}}|0\rangle + \frac{c}{\sqrt{1 + |c|^2}}|1\rangle$$

© Springer-Verlag Berlin Heidelberg 2015
M. Melucci, *Introduction to Information Retrieval and Quantum Mechanics*,
The Information Retrieval Series 35, DOI 10.1007/978-3-662-48313-8

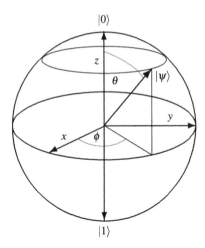

Fig. A.1 The Bloch sphere

When $\alpha_0 = 0$ and $\alpha_1 = 1$, we have that $c = \infty$ by definition. Following this correspondence rule, we have that $|0\rangle$ corresponds to 0, $|1\rangle$ corresponds to ∞, $|i\rangle$ corresponds to i, and $\frac{1}{\sqrt{2}}|0\rangle + \frac{1}{\sqrt{2}}|1\rangle$ corresponds to 1.

Another alternate qubit representation is provided by the Bloch sphere. This sphere is a visual representation of the space within which qubits live. Figure A.1 provides an illustration of the Bloch sphere. The basic idea is that every qubit can be determined by only two angles, that is, θ and ϕ, as any geographical coordinate can be determined by latitude and longitude. This is because any qubit can be a superposition of $|0\rangle$ and $|1\rangle$ where the amplitudes α_0, α_1 are related by

$$|\alpha_0|^2 + |\alpha_1|^2 = 1$$

and therefore, an amplitude is immediately given by the other amplitude. As the amplitudes are complex numbers, it follows that

$$\alpha_0 = \cos\theta$$

and

$$\alpha_1 = e^{i\phi}\sin\theta$$

The north pole of the sphere corresponds to $|0\rangle$ and the south pole corresponds to $|1\rangle$, that is, the poles correspond to the values of the classical bit. Using the Bloch sphere, it is possible to map a complex vector to a complex number as described above and then the complex number $c = (a, b)$ to the real coordinates of the sphere as follows:

$$c \quad \text{corresponds to} \quad \left(\frac{2a}{1+|c|^2}, \frac{2b}{1+|c|^2}, \frac{1-|c|^2}{1+|c|^2} \right)$$

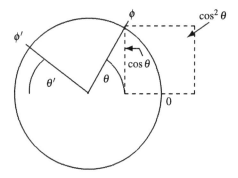

Fig. A.2 Knob and statistical distance

A.2 Why the Trace Rule

There are a couple of rigorous explanations of the use of this rule. One reason is provided by the Wootters statistical distance; the other reason is given by the Gleason theorem.

A.2.1 Wootters' Statistical Distance

The first explanation derives from some arguments presented by Wootters (1981). Imagine a knob for adjusting the relevance grade with which a machine emits documents. When the knob is set to say 0, the machine emits nonrelevant documents; when the knob is set to ϕ, the machine emits documents with relevance grade ϕ. Suppose the knob is a round button labeled by a sign pointing to a relevance grade, thus forming a circle and a tick as depicted in Fig. A.2.

At each ϕ, the tick forms an angle $\theta \in [0, \pi]$ and the machine emits relevant documents with grade ϕ; at a given ϕ, the machine emits documents including the term with probability $p(\theta)$, which might be different from $p(\theta')$ corresponding to a different ϕ'. Therefore, θ is assumed to be related to the probability of term occurrence.

Suppose the occurrences of a given term are counted by observing the documents emitted by the machine. The relative frequency of occurrence gives an estimation of $p(\theta)$ for each position at which the knob has been set. In particular, the relative frequency is the maximum likelihood estimator of the unknown probability; however, there is an estimation error given by the standard deviation of the normal distribution. Indeed, as the number of observations increases, its probability distribution is approximated by the normal distribution with the mean being equal to the relative frequency $\hat{p}$ and the standard deviation being equal to

$$\hat{\sigma}(p) = \sqrt{\frac{\hat{p}(1 - \hat{p})}{n}}$$

Let us replace $\hat{p}$ with $p(\theta)$ and let

$$\sigma(\theta) = \sqrt{\frac{p(\theta)(1-p(\theta))}{n}}$$

be the standard deviation expressed as a function of θ. This function of θ allows us to analyze the uncertainty of θ.

The experiment of observing a term from a document is a Bernoulli trial which is a random variable with two possible outcomes, "term occurs" and "term does not occur," with success having a probability of $p(\theta)$. As for a series of n Bernoulli trials, the probability function is

$$\prod_{i=1}^{n} np(\theta)^{x_i}(1-p(\theta))^{1-x_i}$$

Starting from this probability function, the Fisher information

$$I(\theta) = E\left[(\partial^2 \log f(x;\theta))\right]$$

can be calculated as follows. It can be shown that

$$I(p(\theta)) = \frac{n}{p(\theta)(1-p(\theta))} = \frac{1}{\sigma(\theta)^2} \qquad (A.1)$$

that is, the Fisher information is the reciprocal of the variance. As the Fisher information measures the amount of information that a random variable carries about an unknown parameter p, and this parameter depends on θ, (A.1) measures the amount of information that a random variable carries about θ. Because the Cramér-Rao bound

$$\text{Var}(\hat{\theta}) \geq \frac{1}{I(\theta)}$$

establishes a minimum of the variability of $p(\theta)$, it is possible to set the reciprocal of $I(p(\theta))$ to the smallest variation around $p(\theta)$ when θ is varying. For each θ, this variation is thus fixed, and as a consequence, the Fisher information is fixed too. The variations expressed by an analytical function are described by the derivative of the function. Thus, (A.1) is telling us that the derivative of $p(\theta)$ is proportional to $p(\theta)(1-p(\theta))$, that is,

$$p'(\theta) \propto p(\theta)(1-p(\theta)) \qquad (A.2)$$

The only function of θ such that (A.2) holds is

$$p(\theta) = \cos^2 \theta$$

In other words, whenever the variation of p is explained in terms of a knob which forms an angle θ corresponding to a given relevance grade ϕ, the probability that a term occurs in a document with grade ϕ is modeled by the squared cosine of θ. In particular, if

$$|1\rangle = \begin{pmatrix} 1 \\ 0 \end{pmatrix}$$

is the basis vector of the event that a term occurs placed at angle θ to the state vector $|\phi\rangle$, the probability that the term occurs when the relevance grade is ϕ can be computed as

$$|\langle 1|\phi\rangle|^2 = \cos^2 \theta$$

A.2.2   Gleason's Theorem

The second explanation of the reason why the trace rule is used to compute the probabilities of events represented by vector spaces is given by the Gleason theorem. This theorem is of fundamental importance in QM because it basically states that the density matrix can encapsulate all the information about a probability space, that is, it provides a probability distribution for any conceivable observable. The proof of this simple statement is mathematically very difficult and it is not reported here. What we provide is the statement of the theorem and some comments on its importance; we refer to Hughes (1989) for these comments and further discussion.

Theorem A.1 (Gleason) *A unique density matrix ρ corresponds to every probability distribution on the set of all projectors in a complex vector space with dimension greater than 2 for which the probability of the event represented by every projector $|x\rangle\langle x|$ is* $\mathrm{tr}(\rho|x\rangle\langle x|)$.

The Gleason theorem starts from a complex vector space from which a collection of subspaces can be defined. These subspaces represent propositions, events, answers to questions, or observable values. Whatever they represent, the subspaces are mutually orthogonal, that is, the inner products between the projectors representing these subspaces are null; examples are term occurrence since a term either occurs or not and document relevance since a document is either relevant or not.

The other element of the Gleason theorem is the notion of state. However, the notion of state is viewed in the context of the theorem as a function assigning the value 1 to the whole space, assigning a number in the unit interval to each subspace, and satisfying a simple additivity property, that is, if the projector of any given subspace is written as a sum of the projectors of mutually orthogonal subspaces, then the value of the state on the given "summed" subspace is equal to the sum of the values of the state on the summands. Suppose, for example, that the vector space is defined over the four-dimensional complex space and that a subspace is represented by the following projector

$$|0?\rangle\langle0?| = \begin{pmatrix} 0\ 0\ 0\ 0 \\ 0\ 0\ 0\ 0 \\ 0\ 0\ 1\ 0 \\ 0\ 0\ 0\ 1 \end{pmatrix}$$

Since this projector can be a sum of two projectors as follows

$$|0?\rangle\langle0?| = \begin{pmatrix} 0\ 0\ 0\ 0 \\ 0\ 0\ 0\ 0 \\ 0\ 0\ 1\ 0 \\ 0\ 0\ 0\ 0 \end{pmatrix} + \begin{pmatrix} 0\ 0\ 0\ 0 \\ 0\ 0\ 0\ 0 \\ 0\ 0\ 0\ 0 \\ 0\ 0\ 0\ 1 \end{pmatrix} = |00\rangle\langle00| + |01\rangle\langle01|$$

the value of the state, say p on $|0?\rangle\langle0?|$, is equal to the sum of the values of the state on $|00\rangle\langle00|$ and $|01\rangle\langle01|$, that is,

$$p(|0?\rangle\langle0?|) = p(|00\rangle\langle00|) + p(|01\rangle\langle01|)$$

It is easy to see that we can define a state by associating to each one-dimensional subspace (e.g., $|01\rangle\langle01|$) generated by a unit vector x (e.g., $|01\rangle$) the inner product $\langle x|\rho|x\rangle$ where ρ is a Hermitian matrix with trace 1. States of this type are called regular states.

The key problem addressed by the Gleason theorem was to determine the set of states on an arbitrary complex vector space, that is, to understand the mathematical formulation of the functions assigning a value in the unit interval to a subspace. The problem is not trivial because one expected that a state could possess any mathematical form and that it does not necessarily possess the form given by the trace rule; for example, it might have belonged to an exponential family.

The solution to the problem was given by the Gleason theorem which clarified that every state on a real or complex space of dimension greater than two is necessarily regular, that is, there is no state on such a space with a mathematical formulation other than the product like $\langle x|\rho|x\rangle$, and the only rule to calculate the probabilities is the trace rule, that is,

$$P(x \text{ is observed when state is } \rho) = \text{tr}(\rho|x\rangle\langle x|) = \langle x|\rho|x\rangle$$

A.3   Conditional Probability

In the previous sections, we provided the basics of probability in QM and compared it with classical probability. To this end, we expressed probability distributions, first, as diagonal matrices in the case of classical probability and then as Hermitian matrices in the case of the probability defined within the quantum mechanical framework. In the following paragraphs, we first introduce conditional probability in the classical probability, and we then introduce it in the quantum mechanical framework following the arguments of the book of Aaronson (2013) which can provide the interested reader with further details.

Suppose a bidimensional probability distribution has been arranged in a diagonal density matrix. This density matrix can be viewed as an operator mapping a vector that represents an event to a vector that includes the probability of the event; for example, when the density matrix is

$$\mathbf{P} = \begin{pmatrix} p & 0 \\ 0 & 1 - p \end{pmatrix} \tag{A.3}$$

and the vector that represents an event is

$$\begin{pmatrix} 0 \\ 1 \end{pmatrix}$$

the density matrix acts as an operator mapping this vector to

$$\begin{pmatrix} 0 \\ 1 - p \end{pmatrix}$$

meaning that the event represented by the vector

$$\begin{pmatrix} 0 \\ 1 \end{pmatrix}$$

occurs with probability $1 - p$.

Conditional probability can only be implemented by stochastic matrices, that is, matrices of nonnegative real numbers where every column adds up to unity; for example, the following is a stochastic matrix:

$$\begin{pmatrix} \frac{1}{2} & \frac{1}{3} \\ \frac{1}{2} & \frac{2}{3} \end{pmatrix}$$

By "only be implemented," we mean that a stochastic matrix is the most general matrix that transforms a probability distribution represented by a density matrix into a conditional probability distribution represented by a density matrix. Suppose, for example, the density matrix is (A.3) and

$$\mathbf{Q} = \begin{pmatrix} q & 0 \\ 0 & 1 - q \end{pmatrix}$$

is the resulting conditional density matrix. The following matrix

$$\mathbf{T} = \begin{pmatrix} t_{00} & t_{01} \\ t_{10} & t_{11} \end{pmatrix}$$

such that

$$\mathbf{Q} = \mathbf{TP} \qquad (A.4)$$

is stochastic. Note that $\mathbf{T}$ must be real; if it was complex, $\mathbf{Q}$ would be complex and then would not represent a probability distribution. Moreover, $\mathbf{T}$ must be nonnegative; for example, if t_{00} were negative and $p = 1$, the resulting q would be negative and would then not be a probability distribution. Additionally, the t_{ij}s must be within 0 and 1; indeed, if (A.4) is written as

$$q = t_{00}p + t_{10}(1 - p) \qquad 1 - q = t_{01}p + t_{11}(1 - p) \qquad (A.5)$$

the condition $0 \leq t_{ij} \leq 1$ must hold for each i,j because both p and q are between 0 and 1; if t_{10} were greater than 1 and $p = 0$, q would be greater than 1 and the diagonal of $\mathbf{Q}$ would not be a probability distribution. After summing the expressions of (A.5), we have that

$$1 = (t_{00} + t_{01})p + (t_{10} + t_{11})(1 - p)$$

Since we have argued that $0 \leq t_{ij} \leq 1$, it follows that

$$1 = t_{00} + t_{01} \qquad 1 = t_{10} + t_{11}$$

that is, $\mathbf{T}$ must be stochastic.

To introduce the way conditional probability is described in the quantum mechanical framework, the notion of unit vector and of unitary matrix has to be introduced; actually, the former was already introduced in the preceding sections. A vector $|v\rangle$ is a unit vector when its norm is one, that is, $|\langle v|v \rangle|^2 = 1$.

Unit vectors are crucial because they represent the simplest states from which probability distributions can be obtained for a collection of mutually orthogonal projectors; if a state were not represented as a unitary vector, no probability distribution could be estimated. However, a state may evolve, and therefore, its unit vector may be transformed to another unit vector. It is therefore important to understand what sort of transformation on unit vectors can take place.

A unitary matrix is the most general matrix that transforms a unit vector to another unit vector. A matrix $\mathbf{U}$ is unitary when its transpose equals its inverse, that is,

$$\mathbf{U}^*\mathbf{U} = \mathbf{1}$$

Suppose $\mathbf{U}$ is a matrix that transform a unit vector to another unit vector. In mathematical terms, we have that

$$|u\rangle = \mathbf{U}|v\rangle \qquad |\langle u|u\rangle|^2 = 1$$

The latter expression can be written as

$$|\langle v\mathbf{U}|\mathbf{U}v\rangle|^2 = |\langle v|\mathbf{U}^*\mathbf{U}|v\rangle|^2 = |\langle v|\mathbf{1}|v\rangle|^2 = |\langle v|v\rangle|^2 = 1$$

To show that $\mathbf{U}$ is necessarily unitary, consider the latter expression in terms of column vectors of the matrix, that is,

$$|\langle v|\mathbf{U}^*\mathbf{U}|v\rangle|^2 = (v_1^*\langle u_1| + \cdots + v_n^*\langle u_n|)(v_1|u_1\rangle + \cdots + v_n|u_n\rangle)$$

$$= |v_1|^2 + \cdots + |v_n|^2 + 2\sum_{i=1}^{d}\sum_{j=i+1}^{d} v_i^* v_j^* \langle u_i|u_j\rangle$$

$$= 1 + 2\sum_{i=1}^{d}\sum_{j=i+1}^{d} v_i^* v_j^* \langle u_i|u_j\rangle \tag{A.6}$$

The latter expression equals 1 for any $|v\rangle$ only if $\langle u_i|u_j\rangle = 0$ for each $i \neq j$, that is, $\mathbf{U}$ is unitary. In other terms, if there were two distinct indexes i,j such that $\langle u_i|u_j\rangle \neq 0$, there might be a $|v\rangle$ such that $v_i^* v_j^* \langle u_i|u_j\rangle \neq 0$, thus making (A.6) different from 1. As $|v\rangle$ can arbitrarily be chosen, there cannot be two distinct indexes i,j such that $\langle u_i|u_j\rangle \neq 0$.

A.4 Why Complex Numbers

In IR, it was wondered why QM and then the complex number field are necessary. All things considered, vector spaces have extensively been utilized to define models and design retrieval techniques, and the real number field has shown itself to be sufficient to index and retrieve documents. van Rijsbergen (2004) suggested that complex numbers may be utilized whenever some significant quantity used when indexing and retrieving documents occurs as a pair of real numbers; the TFIDF weighting scheme is an example. However, he also highlighted that it is crucial to distinguish between the use of the real number field in implementing observable values and the use of this field in implementing operators; it is the use of the operators of the quantum mechanical framework that makes the import of QM in IR interesting.

Complex numbers are needed to implement the transformation of a state into a superposed state. To explain this need, it is sufficient to show that there exists a unitary transformation $\mathbf{U}$ such that the solution $\mathbf{V}$ of $\mathbf{U} = \mathbf{V}^2$ must be complex. In

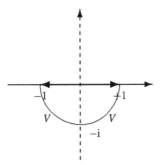

Fig. A.3 Flipping a vector

this case, **U** is a two-step transformation, that is, a transformation performed in two identical steps being each step performed by the application of **V**.

The first step is the transformation represented by the matrix **V**, and the second step is again represented by **V**. Suppose the unitary transformation is

$$\mathbf{U} = \begin{pmatrix} -1 & 0 \\ 0 & 1 \end{pmatrix}$$

and "flips" a vector placed in the one-dimensional line to its opposite located in the same line as depicted in Fig. A.3. This transformation is unitary ($\mathbf{U}^*\mathbf{U} = \mathbf{1}$), since flipping the flipped pure state vector results in the starting vector, i.e.,

$$\mathbf{U}\left(\mathbf{U}\begin{pmatrix} 1 \\ 0 \end{pmatrix}\right) = \mathbf{U}\begin{pmatrix} -1 \\ 0 \end{pmatrix} = \begin{pmatrix} 1 \\ 0 \end{pmatrix}$$

as depicted in Fig. A.3 in which the vector flips to the original position after being flipped. The figure also shows that flipping might be halted in the middle of the travel from one position to another; for example, the vector might be halted in the vertical position oriented toward north in the middle between the two horizontal positions.

To halt flipping in the middle of the travel, it is necessary that a transformation moving the vector from a horizontal position to the vertical position oriented to the north can be defined, that is, something like that moves the vector

$$\begin{pmatrix} 1 \\ 0 \end{pmatrix}$$

to the vector

$$\begin{pmatrix} 0 \\ -i \end{pmatrix}$$

and then moves the latter to the vector

$$\begin{pmatrix} -1 \\ 0 \end{pmatrix}$$

that is, the other horizontal position. However, no real matrix $\mathbf{V}$ exists such that $\mathbf{V}^2 = \mathbf{U}$. Indeed, we have to solve the following equation:

$$\mathbf{V}^2 = \mathbf{U} \qquad \mathbf{V} = \begin{pmatrix} u & v \\ x & y \end{pmatrix}$$

To this end, we have to solve the following system of equations:

$$uu + vx = -1$$
$$uv + vy = 0$$
$$ux + xy = 0$$
$$vx + yy = 1$$

which can be solved only if

$$x = 0 \wedge u = -i \wedge y = -1 \wedge v = 0 \qquad\qquad\text{(A.7)}$$
$$x = 0 \wedge u = -i \wedge y = +1 \wedge v = 0 \qquad\qquad\text{(A.8)}$$
$$x = 0 \wedge u = +i \wedge y = -1 \wedge v = 0$$
$$x = 0 \wedge u = +i \wedge y = +1 \wedge v = 0$$

which is a set of solutions in the complex field since y must be either $-i$ or i, the latter being the imaginary number. Consider the solution (A.8), for example. The "square root" transformation of $\mathbf{U}$, i.e., $\mathbf{V}$ becomes

$$\mathbf{V} = \begin{pmatrix} -i & 0 \\ 0 & 1 \end{pmatrix}$$

The pure state

$$\begin{pmatrix} 1 \\ 0 \end{pmatrix}$$

is transformed by this $\mathbf{V}$ to

$$\begin{pmatrix} -i \\ 0 \end{pmatrix}$$

If **V** is applied again, we have

$$\begin{pmatrix} -1 \\ 0 \end{pmatrix}$$

Visually speaking, this **V** sends

$$\begin{pmatrix} 1 \\ 0 \end{pmatrix}$$

to another dimension provided by the line spanned by i. In this additional, imaginary dimension, the state is no longer a ground state because it is in a superposed state. If this **V** is applied on the superposed state located in the additional dimension, the other ground state

$$\begin{pmatrix} -1 \\ 0 \end{pmatrix}$$

is achieved. This movement is depicted in Fig. A.3.

A.5 No-Cloning Theorem

There is a simple, but important, consequence of the unitary condition imposed on the transformations described in Sect. A.4 when applied to tensored states. The unknown states cannot be cloned. By contrast, the known states can be cloned because the values of their vectors are known with certainty and the cloned states can simply be generated with those values. Consider in the following the proof given by Rieffel and Polak (2011).

Cloning can mathematically be represented by a unitary transformation **U** that takes the state to be cloned and an "empty" state as input and produces the state to be cloned and the cloned state; Fig. A.4 depicts the action of **U**. If it were possible to implement a unitary transformation **U** as a circuit, cloning an unknown state would be possible; the idea of the proof is to show that such a transformation cannot be designed.

As unitary transformations operate on single states, it is necessary to tensor the input states together so that the input state is a single state. Note that the output states are tensored together too since these transformations produce one single state. In mathematical terms, were cloning possible, the transformation should map $|a\rangle \otimes |0\rangle$

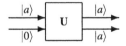

Fig. A.4 Unitary transformation to clone a state

to $|a\rangle \otimes |a\rangle$, that is,

$$|a\rangle \otimes |a\rangle = \mathbf{U}|a\rangle \otimes |0\rangle$$

The unitary matrix acts as a circuit transforming a pair of qubits into another pair of qubits; the question is whether such a circuit can be constructed for any state even if it were a superposition of other states (actually, the no-cloning theorem assures it cannot be). Consider a superposition

$$|\phi\rangle = (|a\rangle + |b\rangle)/\sqrt{2}$$

where $|a\rangle, |b\rangle$ are mutually orthogonal, and the following unitary transformation:

$$\mathbf{U}|\phi\rangle \otimes |0\rangle$$

As unitary transformations are linear, we have that

$$
\begin{aligned}
\mathbf{U}|\phi\rangle \otimes |0\rangle &= \mathbf{U}(|a\rangle + |b\rangle)/\sqrt{2} \otimes |0\rangle \\
&= \mathbf{U}(|a\rangle \otimes |0\rangle + |b\rangle \otimes |0\rangle)/\sqrt{2} \\
&= \mathbf{U}|a\rangle \otimes |0\rangle/\sqrt{2} + \mathbf{U}|b\rangle \otimes |0\rangle/\sqrt{2}
\end{aligned}
$$

If cloning were possible, we should have that

$$\mathbf{U}|a\rangle \otimes |0\rangle = |a\rangle \otimes |a\rangle$$

and that

$$\mathbf{U}|b\rangle \otimes |0\rangle = |b\rangle \otimes |b\rangle$$

and therefore, we should have that

$$\mathbf{U}|\phi\rangle \otimes |0\rangle = |a\rangle \otimes |a\rangle/\sqrt{2} + |b\rangle \otimes |b\rangle/\sqrt{2}$$

and that

$$\mathbf{U}(|\phi\rangle \otimes |0\rangle) = |\phi\rangle \otimes |\phi\rangle$$

However, $|\phi\rangle$ is a superposition, and therefore,

$$
\begin{aligned}
|\phi\rangle \otimes |\phi\rangle &= (|a\rangle + |b\rangle)/\sqrt{2}(|a\rangle + |b\rangle)/\sqrt{2} \\
&= (|a\rangle \otimes |a\rangle + |b\rangle \otimes |a\rangle + |a\rangle \otimes |b\rangle + |b\rangle \otimes |b\rangle)/2
\end{aligned}
$$

which is different from

$$|a\rangle \otimes |a\rangle / \sqrt{2} + |b\rangle \otimes |b\rangle / \sqrt{2}$$

Therefore, cloning using $\mathbf{U}$ is impossible.

A.6 Indeterminacy Principle

The indeterminacy principle is a strange phenomenon of the quantum mechanical framework; another strange phenomenon is entanglement which is described in Sect. 2.6. In physics, the indeterminacy principle is known as the principle that both the momentum and position of a particle cannot be precisely determined at the same time. As mentioned in Sect. 2.2.2, momentum and position are incompatible observables since if the position is measured first and the momentum is measured second, the state of the particle is different from the state obtained if the quantity of motion is measured first and the position is measured second.

The indeterminacy principle is a consequence of incompatibility and differs from incompatibility for its statistical feature; indeed, the indeterminacy principle is a relationship between the variances of two incompatible observables. In particular, the indeterminacy principle states a lower bound of the variance of the conjoint measurement of two incompatible observables and measures the minimum imprecision that must be tolerated when two incompatible observables are conjointly measured; for example, the momentum of a particle can be exactly determined only if the position of the particle can be approximately determined, and vice versa, and the product between the variance of momentum and the variance of precision is always not less than a given lower bound, which is greater than zero when the observables are incompatible.

A.6.1 Observable Operator

In Sect. 2.2, the use of projectors for representing the observable values was illustrated. In short, each observable value x corresponds to a projector $\mathbf{A}_x$; this projector is a mathematical representation of the event that x is observed. The negation of the event that x is observed is represented by a projector orthogonal to $\mathbf{A}_x$ and such that the resolution to unity holds; for example, if $\mathbf{A}_0$ is the projector of the event $x = 0$, $\mathbf{A}_1$ is the projector of the event $x = 1$, and these two events are the only possible events; we have that

$$\mathbf{A}_0 + \mathbf{A}_1 = \mathbf{1} \qquad \mathbf{A}_0 \mathbf{A}_1 = \mathbf{0}$$

A single mathematical representation of an observable can be provided by an operator. This operator is a linear combination of projectors, that is, a weighted sum of projectors where the weights of this combination are the observable values. An operator for an observable can be defined as

$$\mathbf{A} = x_1 \mathbf{A}_1 + \cdots + x_k \mathbf{A}_k$$

where $x_1, \ldots, x_k$ are k observable values referred to the same observable and $\mathbf{A}_1, \ldots, \mathbf{A}_k$ are k projectors that correspond to the k values. The projectors are mutually orthogonal and resolve to the unity, that is,

$$\mathbf{A}_0 + \cdots + \mathbf{A}_k = \mathbf{1} \qquad \mathbf{A}_i \mathbf{A}_j = \mathbf{0} \quad i \neq j \qquad \mathbf{A}_i \mathbf{A}_i = \mathbf{A}_i \quad \text{for all } i$$

This representation is useful since the calculation of the observable value that corresponds to a projector can be performed as follows. Consider the projector $\mathbf{A}_i$. The observable value can be extracted by the following expression:

$$x_i = \text{tr}(\mathbf{A}\mathbf{A}_i)$$

since

$$
\begin{aligned}
\text{tr}(\mathbf{A}\mathbf{A}_i) &= \text{tr}((x_1 \mathbf{A}_1 + \cdots + x_k \mathbf{A}_k)\mathbf{A}_i) \\
&= \text{tr}(x_1 \mathbf{A}_1 \mathbf{A}_i + \cdots + x_k \mathbf{A}_k \mathbf{A}_i) \\
&= \text{tr}(x_i \mathbf{A}_i \mathbf{A}_i) \\
&= x_i \text{tr}(\mathbf{A}_i) \\
&= x_i
\end{aligned}
$$

Recall that a density matrix can be defined for the observable represented by $\mathbf{A}$; for example, suppose a pure distribution can be defined and represented by the state vector $|\phi\rangle$ or equivalently by the pure density matrix

$$\rho = |\phi\rangle\langle\phi|$$

It follows that

$$\rho = p_1 \mathbf{A}_1 + \cdots + p_k \mathbf{A}_k$$

where

$$p_1 + \cdots + p_k = 1 \quad \text{and} \quad 0 \leq p_i \leq 1 \quad i = 1, \ldots, k$$

As explained in Sect. 2.4, the probabilities can be computed using the trace rule, that is,

$$p_i = \text{tr}(\rho \mathbf{A}_i) \quad i = 1, \ldots, k$$

For example, suppose $k = 2$ and let $\mathbf{A}$ be the operator of the observable defined for the occurrence of a term in a document. Let $x = 0$ be the observable value when the term does not occur, and let $x = 1$ be the observable value when the term occurs. With probability p_0, the term does not occur, and with probability $p_1 = 1 - p_0$, the term occurs. We have that

$$\mathbf{A} = \mathbf{A}_1$$

where $\mathbf{A}_1$ is the projector of the event that the term occurs. Note that there are other operators that can be defined for the same observable; for example, let $x = -1$ be the observable value when the term does not occur, and let $x = +1$ be the observable value when the term occurs; we have that

$$\mathbf{A}_1 - \mathbf{A}_0$$

is another operator.

A.6.2 Expectation and Variance

Expectation and variance can be calculated using the observable operator and the density matrix. The expectation of a random variable that takes the values $x_1, \ldots, x_k$ according to the probability distribution $p_1, \ldots, p_k$ is defined as

$$E = x_1 p_1 + \cdots + x_k p_k$$

whereas the variance of the random variable is defined as

$$V = (x_1 - E)^2 p_1 + \cdots + (x_k - E)^2 p_k$$

An observable represented by the operator $\mathbf{A}$ corresponds to a random variable, and the expectation of the observable can be defined as

$$E(\mathbf{A}) = E$$

The probabilities can be expressed using the trace rule as follows:

$$E(\mathbf{A}) = x_1 \text{tr}(\rho \mathbf{A}_1) + \cdots + x_k \text{tr}(\rho \mathbf{A}_k)$$

Using the properties of linear combination, the following expression is obtained:

$$E(\mathbf{A}) = \mathrm{tr}(\rho\mathbf{A})$$

Using similar arguments, the variance of the observable can be defined as follows:

$$V(\mathbf{A}) = \mathrm{tr}(\rho\mathbf{A}^2) - \mathrm{tr}(\rho\mathbf{A})^2$$

where

$$\begin{aligned}
\mathbf{A}^2 &= \mathbf{A}\mathbf{A} \\
&= (x_1\mathbf{A}_1 + \cdots + x_k\mathbf{A}_k)(x_1\mathbf{A}_1 + \cdots + x_k\mathbf{A}_k) \\
&= x_1^2\mathbf{A}_1 + \cdots + x_k^2\mathbf{A}_k
\end{aligned}$$

A.6.3   Uncertainty Inequality

Recall that the conjoint measurement of two observables that are represented by two operators can be expressed as the following matrix product:

$$\mathbf{A}\mathbf{B}$$

where

$$\mathbf{B} = y_1\mathbf{B}_1 + \cdots + y_m\mathbf{B}_m$$

is the operator of another observable, the y_js are the values of this observable, and the $\mathbf{B}_j$s are the projectors of the event that the y_js are observed. Expectation and variance can also be computed for this additional observable. In general, two observables are incompatible, and therefore, their operators do not commute, that is,

$$\mathbf{A}\mathbf{B} \neq \mathbf{B}\mathbf{A}$$

It can be shown that two observable operators commute when each pair of projectors commutes, that is,

$$\mathbf{A}_i\mathbf{B}_j = \mathbf{B}_j\mathbf{A}_i \qquad i = 1,\ldots,k \qquad j = 1,\ldots,m$$

Therefore, if two observables are not compatible in the real world, their mathematical representation should be provided by two noncommutative operators; for example, the following operators do not commute:

$$x_0 \begin{pmatrix} 0 & 0 \\ 0 & 1 \end{pmatrix} + x_1 \begin{pmatrix} 1 & 0 \\ 0 & 0 \end{pmatrix}$$

$$\frac{y_0}{2} \begin{pmatrix} 1 & -1 \\ -1 & 1 \end{pmatrix} + \frac{y_1}{2} \begin{pmatrix} 1 & 1 \\ 1 & 1 \end{pmatrix}$$

whereas the following operators commute:

$$x_0 \begin{pmatrix} 0 & 0 & 0 \\ 0 & 0 & 0 \\ 0 & 0 & 1 \end{pmatrix} + x_1 \begin{pmatrix} 1 & 0 & 0 \\ 0 & 1 & 0 \\ 0 & 0 & 0 \end{pmatrix}$$

$$y_0 \begin{pmatrix} 0 & 0 & 0 \\ 0 & 0 & 0 \\ 0 & 0 & 1 \end{pmatrix} + y_1 \begin{pmatrix} 0 & 0 & 0 \\ 0 & 1 & 0 \\ 0 & 0 & 0 \end{pmatrix} + y_2 \begin{pmatrix} 1 & 0 & 0 \\ 0 & 0 & 0 \\ 0 & 0 & 0 \end{pmatrix}$$

Consider a special observable called commutator. A commutator does not correspond to a physical measurement device; it is only a function of two observables, and it is a mathematical expression of the conjoint measurement performed through two observables. A commutator is defined so as to yield 0 when two observables are compatible and a value different from 0 when two observables are incompatible. The operator of the commutator of two observables is defined as

$$\mathbf{C} = \mathbf{AB} - \mathbf{BA}$$

In the remainder of this section, we follow the arguments provided by Nielsen and Chuang (2000). Suppose two observable operators are defined with null expectation; if the corresponding observables do not exhibit null expectation, it is always possible to scale the observable values so as to obtain null expectation; for example, the expectation of the observable that corresponds to the occurrence of a term and that has values 0 and 1 with uniform probability $\frac{1}{2}$ is $\frac{1}{2}$; an observable that has null expectation can be obtained by subtracting the expectation from each observable value. The variance of the operator of an observable with null expectation is defined as

$$V(\mathbf{A}) = \mathrm{tr}(\rho \mathbf{A}^2)$$

Consider the product between the variances of two observables with null expectation:

$$V(\mathbf{A})V(\mathbf{B}) = \mathrm{tr}(\rho \mathbf{A}^2)\mathrm{tr}(\rho \mathbf{B}^2) = \langle \phi | \mathbf{A}^2 | \phi \rangle \langle \phi | \mathbf{B}^2 | \phi \rangle$$

Define the following vectors

$$|\psi\rangle = \mathbf{A}|\phi\rangle \qquad |\xi\rangle = \mathbf{B}|\phi\rangle$$

Note that

$$\langle\phi|\mathbf{A}^2|\phi\rangle = \langle\phi|\mathbf{AA}|\phi\rangle = ((\langle\phi|\mathbf{A})(\mathbf{A}|\phi\rangle)) = \langle\psi|\psi\rangle$$

and

$$\langle\phi|\mathbf{B}^2|\phi\rangle = \langle\phi|\mathbf{BB}|\phi\rangle = ((\langle\phi|\mathbf{B})(\mathbf{B}|\phi\rangle)) = \langle\xi|\xi\rangle$$

Consider the Cauchy-Schwartz inequality that states that

$$\langle\psi|\psi\rangle\langle\xi|\xi\rangle \geq \langle\psi|\xi\rangle\langle\xi|\psi\rangle$$

After replacing $|\psi\rangle$ and $|\xi\rangle$, the following inequality is obtained:

$$V(\mathbf{A})V(\mathbf{B}) \geq |\langle\phi|\mathbf{AB}|\phi\rangle|^2$$

Note that

$$\langle\phi|\mathbf{AB}|\phi\rangle$$

is a complex number $a + ib$ such that

$$a = \frac{1}{2}\langle\phi|\mathbf{AB} + \mathbf{BA}|\phi\rangle \qquad b = \frac{1}{2}\langle\phi|\mathbf{AB} - \mathbf{BA}|\phi\rangle$$

It follows that

$$|\langle\phi|\mathbf{AB}|\phi\rangle|^2 = a^2 + b^2$$

which is never negative, and in particular, it is never less than b^2. However,

$$b^2 = \frac{1}{4}|\langle\phi|\mathbf{C}|\phi\rangle|^2$$

thus obtaining the following inequality:

$$V(\mathbf{A})V(\mathbf{B}) \geq \frac{1}{4}|\langle\phi|\mathbf{C}|\phi\rangle|^2 \tag{A.9}$$

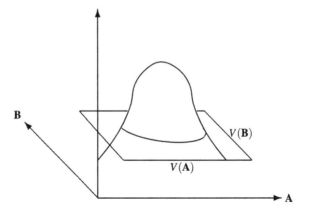

Fig. A.5 Indeterminacy principle

which quantifies the indeterminacy principle also known as Heisenberg's uncertainty principle. Figure A.5 depicts an illustration of this principle.[1] The tridimensional plot represents the probability of the conjoint measurement of the observables represented by the operators **A** and **B**. The left-hand side of (A.9) is the measure of the area of the plane cutting the plot. The inequality says that the shape of the plot is constrained by the right-hand side of (A.9) in a way that the volume below the plot cannot be reduced and that only the shape can be "squeezed," thus increasing $V(\mathbf{A})$ while reducing $V(\mathbf{B})$ or vice versa. Squeezing the plot is an intuitive description of what happens to the precision of the measurement of one observable when the precision of the measurement of the other observable is changed: the reduction of the latter causes the increase of the former. This inverse relationship between the precisions and the inequality (A.9) means that there is an amount of indeterminacy which is represented by the volume below the plot and cannot be eliminated.

Consider the following operators to show how the incompatibility between relevance and aboutness causes the Heisenberg uncertainty inequality (see also Sect. 3.2.6). Suppose operator **A** represents aboutness and operator **R** represents relevance. A possible implementation can be based on the following values and projectors:

$$\mathbf{A} = 0\mathbf{A}_0 + 1\mathbf{A}_1 \qquad \mathbf{B} = 0\mathbf{B}_0 + 1\mathbf{B}_1$$

where

$$\mathbf{A}_1 = \begin{pmatrix} 0 & 1 \\ 1 & 0 \end{pmatrix} \qquad \mathbf{B}_1 = \begin{pmatrix} 0 & -1 \\ 1 & 0 \end{pmatrix}$$

---

[1]The figure was inspired by Maccone and Salasnich (2008).

Suppose a document ϕ can be represented by a state vector $|\phi\rangle$; this state vector determines the probability distribution of the observables and, for example, is defined as follows:

$$|\phi\rangle = \begin{pmatrix} 1 \\ 0 \end{pmatrix}$$

The commutator of relevance and aboutness is

$$\mathbf{C} = \begin{pmatrix} 2 & 0 \\ 0 & -2 \end{pmatrix}$$

Therefore, the right-hand side of (A.9) is

$$\frac{1}{4}|\langle\phi|\mathbf{C}|\phi\rangle|^2 = 1$$

and we have that

$$V(\mathbf{A})V(\mathbf{B}) \geq 1$$

Note that the choice of the operators that implement the observables is crucial. Indeed, if the following matrices were chosen

$$\mathbf{A}_1 = \begin{pmatrix} 1 & 0 \\ 0 & 0 \end{pmatrix} \qquad \mathbf{B}_1 = \begin{pmatrix} \frac{1}{2} & \frac{1}{2} \\ \frac{1}{2} & \frac{1}{2} \end{pmatrix}$$

the right-hand side of (A.9) would always be 0 for all state vectors even if these two matrices represent incompatible observables. Note also that the choice of the state vector is crucial. Indeed, if the state vector were

$$\frac{1}{\sqrt{2}} \begin{pmatrix} 1 \\ 1 \end{pmatrix}$$

the right-hand side of (A.9) would be 0 even if the commutator is $\mathbf{C}$.

A.7 Bell's Inequality

In this section, the origin and the violation of Bell's inequality mentioned in Sect. 2.6.2 are briefly explained. We utilize an IR-like language to make the explanation easier to follow; in Sect. 3.12, some more technical readings are suggested.

Some users are asked to assess some features of documents. The documents are prepared as pairs, that is, one pair of documents are given to the users at a time. In particular, the first document of a pair is given to one user, and the second document is given to the other user. The users are supposed to be faraway so that they cannot communicate. The users are asked to measure one feature at a time; for example, one user is asked to measure relevance and the other user has to measure aboutness. Each user may change the observable at each document; he may measure aboutness of a document and measure term occurrence of the next document; sometimes, they measure the same observable, and sometimes they do not. When observing the sequence of documents sent to a user, two main events occur: the user selects the observable (i.e., aboutness, relevance, occurrence) and the user measures the selected observable (e.g., if relevance was selected, the user measures either relevant or irrelevant).

When the sequence of documents sent to the users ends, three frequency tables can be prepared, one table for each pair of observables:

	occurs	does not occur
relevant	$P(R = 1, X = 1)$	$P(R = 1, X = 0)$
irrelevant	$P(R = 0, X = 1)$	$P(R = 0, X = 0)$

	occurs	does not occur
about	$P(A = 1, X = 1)$	$P(A = 1, X = 0)$
not about	$P(A = 0, X = 1)$	$P(A = 0, X = 0)$

	about	not about
relevant	$P(R = 1, A = 1)$	$P(R = 1, A = 0)$
irrelevant	$P(R = 0, A = 1)$	$P(R = 0, A = 0)$

where R is the observable (i.e., random variable) of relevance such that $R = 1$ means "relevant" and $R = 0$ otherwise, whereas A is the observable of aboutness and X is the observable of term occurrence. The number $P(R = 1, A = 1)$ is the probability (i.e., relative frequency) that the first document of a pair was relevant, and the second document of the same pair was about a given topic. It can be checked that the sum of the four probabilities of each table must be one. Moreover, the marginal probabilities of one table should match with the marginal probabilities of another table within a "small" statistical error, in particular:

$$P(R = 1) = P(R = 1, A = 0) + P(R = 1, A = 1) = P(R = 1, X = 0) + P(R = 1, X = 1)$$

Bell (1964) also noted the following inequality:

$$P(R = 1, X = 0) + P(R = 0, X = 1) + P(R = 1, A = 0) +$$
$$P(R = 0, A = 1) + P(X = 1, A = 0) + P(X = 0, A = 1) \leq 2 \quad \text{(A.10)}$$

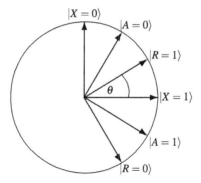

Fig. A.6 Observable polarization in the vector space

This inequality is valid for all triples of observables. However, it is violated when *entangled* photons replace documents and relevance; aboutness and occurrence are replaced by the measurement of three different polarizations, for example, the polarization at angle 0, the polarization at angle 30 degrees, and the polarization at angle 60 degrees. It was found that the mathematical model that explains the experimental results obtained from entangled photons and polarizations is superposition as follows:

$$|R = 1\rangle = \cos\theta|X = 1\rangle + \sin\theta|X = 0\rangle$$

where

$$P(R = 1, X = 1) = |\langle R = 1|X = 1\rangle|^2 = \cos^2\theta$$

and

$$P(R = 1, X = 0) = |\langle R = 1|X = 0\rangle|^2 = \sin^2\theta$$

and $|R = 1\rangle$ is the state vector of the first document when it is relevant (i.e., the event that the first document is relevant), $|X = 1\rangle$ is the event that the term occurs in the second document, and θ is the angle between polarizations; we are indeed viewing the relationship between relevance, aboutness, and occurrence as the geometrical relationship between three polarizations placed at angle θ; this relationship is visualized in Fig. A.6. As relevance and irrelevance are mutually exclusive, the following vector must be orthogonal to $|R = 1\rangle$:

$$|R = 0\rangle = \sin\theta|X = 1\rangle - \cos\theta|X = 0\rangle$$

Moreover,

$$|A = 1\rangle = \cos\theta|X = 1\rangle - \sin\theta|X = 0\rangle$$

and

$$|A = 0\rangle = \sin \theta |X = 1\rangle + \cos \theta |X = 0\rangle$$

The probabilities of the first two tables can then be updated as follows:

	occurs	does not occur
relevant	$\cos^2 \theta$	$\sin^2 \theta$
irrelevant	$\sin^2 \theta$	$\cos^2 \theta$

	occurs	does not occur
about	$\cos^2 \theta$	$\sin^2 \theta$
not about	$\sin^2 \theta$	$\cos^2 \theta$

The probabilities of the third table are computed as follows:

$$
\begin{aligned}
P(R = 1, A = 0) &= |\langle R = 1|A = 0\rangle|^2 \\
&= |\langle R = 1|(\sin \theta |X = 1\rangle + \cos \theta |X = 0\rangle)|^2 \\
&= |\sin \theta \langle R = 1|X = 1\rangle + \cos \theta \langle R = 1|X = 0\rangle|^2 \\
&= |\sin \theta \cos \theta + \cos \theta \sin \theta|^2 \\
&= |2 \sin \theta \cos \theta|^2 \\
&= 4 \sin^2 \theta \cos^2 \theta
\end{aligned}
$$

$$
\begin{aligned}
P(R = 0, A = 1) &= |\langle R = 0|A = 1\rangle|^2 \\
&= |\langle R = 0|(\cos \theta |X = 1\rangle - \sin \theta |X = 0\rangle)|^2 \\
&= |\cos \theta \langle R = 0|X = 1\rangle - \sin \theta \langle R = 0|X = 0\rangle|^2 \\
&= |\cos \theta \sin \theta + \sin \theta \cos \theta|^2 \\
&= |2 \sin \theta \cos \theta|^2 \\
&= 4 \sin^2 \cos^2 \theta
\end{aligned}
$$

The probabilities that are reexpressed as functions of θ can be plugged in Bell's inequality (A.10) to obtain the following expression:

$$4(2 + \cos 2\theta) \sin^2 \theta \le 2 \tag{A.11}$$

The left-hand side of (A.11) is depicted in Fig. A.7 which shows when the inequality can be violated.

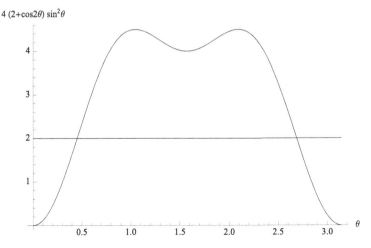

$4 (2+\cos2\theta) \sin^2\theta$

Fig. A.7 Bell's inequality

A.8 The Schmidt Number and Decomposition

This decomposition exploits the SVD and provides an algorithm for expressing a state vector of a product space as a superposition of pure state vectors, thus assuring that a basis exists for each component system.

Suppose A, B are two n_A-dimensional and n_B-dimensional spaces, respectively; for example, A may refer to the occurrence of one term, and an orthonormal basis of this space can represent the event that either the term occurs or does not ($n_A = 2$); similarly, B may refer to the occurrence of another term ($n_B = 2$).

The theorem of the Schmidt decomposition states that for any state $|\phi\rangle$ of the product space $A \otimes B$, there exist orthonormal basis states for A and orthonormal basis states for B such that $|\phi\rangle$ is a superposition of the product basis states, that is,[2]

$$|\phi\rangle = \sum_{i=1}^{\min n_A, n_B} \lambda_{i,i} |i_A\rangle \otimes |i_B\rangle \qquad \sum_i \lambda_{i,i}^2 = 1$$

where

$$\lambda_{i,i} \in \mathbb{R} \qquad |\phi\rangle \in A \otimes B \qquad |i_A\rangle \in A \qquad |i_B\rangle \in B$$

This result can be proved after observing that a state vector $|\phi\rangle \in A \otimes B$ can be expressed as a linear combination of an arbitrary orthonormal basis of the product space $|j\rangle \otimes |k\rangle$, where $|j\rangle$ and $|k\rangle$ are two arbitrary orthonormal bases of A and B,

---

[2]The finite-dimensional case is assumed for the sake of simplicity.

respectively, that is,

$$|\phi\rangle = \sum_{j,k} c_{j,k} |j\rangle \otimes |k\rangle,$$

where the $c_{j,k}$ are the coefficients of the linear combination. These coefficients can be arranged as a $n_B \times n_A$ matrix $\mathbf{C}$, thus establishing a connection to the SVD. Indeed, this latter result states that there exist three matrices $\mathbf{B}, \Lambda, \mathbf{A}$ such that

$$\mathbf{C} = \mathbf{B}\Lambda\mathbf{A}$$

where $\mathbf{B}$ is $n_B \times n_B$, Λ is $n_B \times n_A$ such that $\lambda_{i,j} = 0$ when $i \neq j$, and $\mathbf{A}$ is $n_A \times n_A$, that is,

$$c_{j,k} = \sum_i b_{j,i} \lambda_{i,i} a_{i,k} .$$

After substituting,

$$|\phi\rangle = \sum_{j,k,i} b_{j,i} \lambda_{i,i} a_{i,k} |j\rangle \otimes |k\rangle$$

$$= \sum_i \lambda_{i,i} |i_A\rangle \otimes |i_B\rangle$$

where

$$|i_A\rangle = \sum_j b_{j,i} |j\rangle \qquad |i_B\rangle = \sum_k a_{i,k} |k\rangle$$

Consider, for example,

$$|\phi_1\rangle = \frac{1}{\sqrt{2}} \begin{pmatrix} 1 \\ 0 \\ 0 \\ 1 \end{pmatrix}$$

which can be rewritten as

$$\mathbf{C} = \frac{1}{\sqrt{2}} \begin{pmatrix} 1 & 0 \\ 0 & 1 \end{pmatrix}$$

We have that $n_A = n_B = 2$ and that

$$\mathbf{C} = \frac{1}{\sqrt{2}} \begin{pmatrix} 1 & 0 \\ 0 & 1 \end{pmatrix} = \begin{pmatrix} 1 & 0 \\ 0 & 1 \end{pmatrix} \begin{pmatrix} \frac{1}{\sqrt{2}} & 0 \\ 0 & \frac{1}{\sqrt{2}} \end{pmatrix} \begin{pmatrix} 1 & 0 \\ 0 & 1 \end{pmatrix}$$

The Schmidt number is 2, and then $|\phi_1\rangle$ is entangled. Moreover, consider the following state vector:

$$\frac{1}{2}(|00\rangle + |01\rangle + |10\rangle + |11\rangle)$$

We have that $n_A = n_B = 2$ and that

$$\mathbf{C} = \frac{1}{2}\begin{pmatrix} 1 & 1 \\ 1 & 1 \end{pmatrix} = \frac{1}{2}\begin{pmatrix} 1 & 1 \\ -1 & 1 \end{pmatrix}\begin{pmatrix} 1 & 0 \\ 0 & 0 \end{pmatrix}\begin{pmatrix} 1 & 1 \\ -1 & 1 \end{pmatrix}$$

The Schmidt number is 1, and then this state vector is not entangled.

References

S. Aaronson. *Quantum Computing Since Democritus*. Cambridge University Press, 2013.

L. Accardi. On the probabilistic roots of the quantum mechanical paradoxes. In S. Diner and L. de Broglie, editors, *The Wave-Particle Dualism*, pages 297–330. D. Reidel pub. co., 1984.

L. Accardi. *Urne e camaleonti*. Il Saggiatore, 1997.

L. Accardi and A. Fedullo. On the statistical meaning of complex numbers in quantum mechanics. *Lettere al nuovo cimento*, 34(7):161–172, 1982.

D. Aerts. Quantum structure in cognition. *Journal of Mathematical Psychology*, 53(5):314–348, 2009.

D. Aerts and M. Czachor. Quantum aspects of semantic analysis and symbolic artificial intelligence. *Journal of Physics A: Mathematical and General*, 37(12):L123, 2004. URL http://stacks.iop.org/0305-4470/37/i=12/a=L01.

D. Aerts and L. Gabora. A theory of concepts and their combinations: I.: The structure of the sets of contexts and properties. *Kybernetes*, 34:(1/2):167–191, 2004a.

D. Aerts and L. Gabora. A theory of concepts and their combinations: II.: A Hilbert space representation. *Kibernetes*, 34(1/2):192–221, 2004b.

R. Agrawal, H. Mannila, R. Srikant, H. Toivonen, and A. I. Verkamo. Fast discovery of association rules. In U. M. Fayyad, G. Piatetsky-Shapiro, P. Smyth, and R. Uthurusamy, editors, *Advances in Knowledge Discovery and Data Mining*, pages 307–328. AAAI/MIT Press, 1996.

D. Albert. *Quantum Mechanics and Experience*. Harvard University Press, 1994.

O. E. Barndorff-Nielsen, R. D. Gill, and P. E. Jupp. On quantum statistical inference. *Journal of the Royal Statistical Society. Series B (Statistical Methodology)*, 65(4):775–816, 2003.

D. Bartholomew, F. Steele, and I. Moustaki. *Analysis of multivariate social science data*. Statistics in the social and behavioral sciences series. CRC Press, 2008.

J. Bell. On the Einstein Podolsky Rosen paradox. *Physics*, pages 195–200, 1964.

J. Bell. *Speakable and Unspeakable in Quantum Mechanics*. Cambridge University Press, 1987.

G. Boole. *An Investigation of the laws of Thought*. Walton and Maberly, 1854.

S. Brin and L. Page. The anatomy of a large-scale hypertextual web search engine. In *Proceedings of WWW*, Brisbane, Australia, 1998. http://www7.scu.edu.au/.

P. Bruza and R. Cole. *Quantum Logic of Semantic Space: An Exploratory Investigation of Context Effects in Practical Reasoning*, volume 1, pages 339–362. College Publications, 2005.

P. Bruza, K. Kitto, D. Nelson, and C. McEvoy. Is there something quantum-like about the human mental lexicon? *Journal of Mathematical Psychology*, 53:362–377, 2010.

© Springer-Verlag Berlin Heidelberg 2015

M. Melucci, *Introduction to Information Retrieval and Quantum Mechanics*,
The Information Retrieval Series 35, DOI 10.1007/978-3-662-48313-8

P. Bruza, K. Kitto, B. Ramm, L. Sitbon, S. Blomberg, and D. Song. Quantum-like non-separability of concept combinations, emergent associates and abduction. *Logic Journal of the IGPL*, 20(2):445–457, 2012.

J. Busemeyer and P. D. Bruza. *Quantum Models of Cognition and Decision*. Cambridge University Press, 2012.

G. Cariolaro. *Quantum Communications*. Springer, 2015.

C. Cleverdon and J. Mills. The testing of index language devices. *ASLIB Proceedings*, 15(4): 106–130, 1963.

C. Cleverdon, J. Mills, and M. Keen. *ASLIB Cranfield Research Project: factors determining the performance of indexing systems*. ASLIB, 1966.

W. Cooper. Getting beyond Boole. *Information Processing & Management*, 24:243–248, 1988.

W. Cooper. Some inconsistencies and misidentified modeling assumptions in probabilistic information retrieval. *ACM Transactions on Information Systems*, 13(1):100–111, 1995.

W. Croft and D. Harper. Using probabilistic models of document retrieval without relevance information. *Journal of Documentation*, 35:285–295, 1979.

W. Croft and J. Lafferty, editors. *Language Modeling for Information Retrieval*, volume 13 of *Kluwer International Series on Information Retrieval*. Kluwer Academic Publishers, 2002.

W. Croft, D. Metzler, and T. Strohman. *Search Engines: Information Retrieval in Practice*. Addison Wesley, 2009.

S. Deerwester, S. Dumais, G. Furnas, T. Landauer, and R. Harshman. Indexing by latent semantic analysis. *Journal of the American Society for Information Science*, 41(6):391–407, 1990.

P. Dirac. *The Principles of Quantum Mechanics*. Clarendon Press, 1935.

D. Dubin. The most influential paper Gerard Salton never wrote. *Library Trends*, 52(4):748–764, 2004.

A. Einstein, B. Podolski, and N. Rosen. Can quantum-mechanical description of physical reality be considered complete? *Physical Review*, 47, 1935.

R. Feynman, R. Leighton, and M. Sands. *The Feynman Lectures On Physics*. Addison-Wesley, 1965.

A. Fine. Probability and the interpretation of quantum mechanics. *The British Journal for the Philosophy of Science*, 24(1):1–37, 1973.

E. Fox. Characterization of two new experimental collections in computer and information science containing textual and bibliographic concepts. Technical Report TR83-561, Cornell University, Computer Science Department, 1983.

W. Frakes and R. Baeza-Yates, editors. *Information Retrieval: Data Structures and Algorithms*. Prentice Hall, 1992.

I. Frommholz, B. Larsen, B. Piwowarski, M. Lalmas, P. Ingwersen, and C. J. van Rijsbergen. Supporting polyrepresentation in a quantum-inspired geometrical retrieval framework. In *Proceedings of IIiX*, pages 115–124, 2010.

N. Fuhr. A probability ranking principle for interactive information retrieval. *Journal of Information Retrieval*, 11(3):251–265, 2008.

P. Gärdenfors. *Conceptual Spaces: The Geometry of Thought*. MIT Press, 2000.

A. M. Gleason. Measures on the closed subspaces of a Hilbert space. *Journal of Mathematics and Mechanics*, 6:885–893, 1957.

R. B. Griffiths. *Consistent Quantum Theory*. Cambridge University Press, 2002.

P. Halmos. *Finite-Dimensional Vector Spaces*. Undergraduate Texts in Mathematics. Springer, 1987.

J. Han, J. Pei, and Y. Yin. Mining frequent patterns without candidate generation. In *Proceedings of SIGMOD*, pages 1–12, 2000.

G. M. Hardegree. An approach to the logic of natural kinds. *Pacific Philosophical Quarterly*, 63: 122–132, 1982.

L. Hardy. Quantum theory from five reasonable axioms. http://arxiv.org/abs/quant-ph/0101012v4, 2001.

D. Harman. Relevance feedback revisited. In *Proceedings of SIGIR*, pages 1–10, Copenhagen, 1992.

C. Helstrom. *Quantum Detection and Estimation Theory*. Academic Press, 1976.

W. Hersh, C. Buckley, T. Leone, and D. Hickam. OHSUMED: an interactive retrieval evaluation and new large test collection for research. In *Proceedings of SIGIR*, pages 192–201, Dublin, Ireland, 1994.

Y. Hou and D. Song. Characterizing pure high-order entanglements in lexical semantic spaces via information geometry. In *Proceedings of Quantum Interaction*, pages 237–250, 2009.

Y. Hou, X. Zhao, D. Song, and W. Li. Mining pure high-order word associations via information geometry for information retrieval. *ACM Transactions on Information Systems*, 31(3):1–32, 2013.

R. Hughes. *The Structure and Interpretation of Quantum Mechanics*. Harvard University Press, 1989.

P. Ingwersen. *Information Retrieval Interaction*. Taylor Graham Publishing, 1992.

P. Ingwersen and K. Järvelin. *The Turn: Integration of Information Seeking and Retrieval in Context*. Springer, 2005.

N. Jardine and C. J. van Rijsbergen. The use of hierarchical clustering in information retrieval. *Information Storage and Retrieval*, 7:217–240, 1971.

J. M. Jauch. *Foundations of Quantum Mechanics*. Addison Wesley, 1968.

D. Kelly. *Understanding Implicit Feedback and Document Preference: A Naturalistic User Study*. PhD thesis, Rutgers, The State University of New Jersey, 2004.

D. Kelly. Methods for evaluating interactive information retrieval systems with users. *Foundations and Trends in Information Retrieval*, 3(1-2):1–224, 2009.

A. Khrennikov. *Ubiquitous Quantum Structure*. Springer, 2010.

A. Kolmogorov. *Foundations of the Theory of Probability*. Chelsea Publishing Company, II edition, 1956.

M. Kumar. *Quantum: Einstein, Bohr, and the Great Debate About the Nature of Reality*. W.W. Norton and Co., 2008.

J. Lafferty and C. Zhai. *Probabilistic relevance models based on document and query generation*, chapter 1. Volume 13 of Croft and Lafferty (2002), 2002.

V. Lavrenko and W. Croft. Relevance-based language models. In *Proceedings of SIGIR*, pages 120–127, New Orleans, LO, USA, 2001.

L. M. Lederman and C. T. Hill. *Quantum Physics for Poets*. Prometheus Books, 2011.

T.-Y. Liu. *Learning to Rank for Information Retrieval*. Springer, 2011.

J. Lovins. Development of a stemming algorithm. *Mechanical Translation and Computational Linguistics*, 11:22–31, 1968.

H. Luhn. The automatic creation of literature abstracts. *IBM Journal of Research and Development*, 2(2):159–165, 1958.

H. Luhn. Keyword-in-context index for technical literature. *American Documentation*, 11(4): 288–294, 1960.

L. Maccone and L. Salasnich. *Meccanica quantistica, caos e sistemi complessi*. Carocci Editore, 2008.

J. D. Malley and J. Hornstein. Quantum statistical inference. *Statistical Science*, 8(4):433–457, 1993.

C. Manning and H. Schütze. *Foundations of Statistical Natural Language Processing*. The MIT Press, 1999.

C. Manning, P. Raghavan, and H. Schütze. *An Introduction to Information Retrieval*. Cambridge University Press, 2008.

M. Maron and J. Kuhns. On relevance, probabilistic indexing and retrieval. *Journal of the ACM*, 7:216–244, 1960.

Z. Meglicki. *Quantum Computing without Magic*. Cambridge University Press, 2008.

M. Melucci. A basis for information retrieval in context. *ACM Transactions on Information Systems*, 26(3), 2008a.

M. Melucci. Towards modeling implicit feedback with quantum entanglement. In *Proceedings of Quantum Interaction* Oxford, UK, 2008b. College Publications.

M. Melucci. Deriving a quantum information retrieval basis. *The Computer Journal*, 56(11): 1279–1291, 2012a.

M. Melucci. *Contextual Search: A Computational Framework*. Foundations and Trends in Information Retrieval. Now Publishers, 2012b.

M. Melucci and R. White. Utilizing a geometry of context for enhanced implicit feedback. In *Proceedings of CIKM*, pages 273–282, New York, NY, USA, 2007. ACM.

N. D. Mermin. *Quantum Computer Science: An Introduction*. Cambridge University Press, 2007.

C. Mooers. Coding, information retrieval, and the rapid selector. *Journal of Documentation*, 1(4): 225–229, 1950.

J. Neyman and E. Pearson. On the problem of the most efficient tests of statistical hypotheses. *Philosophical Transactions of the Royal Society, Series A*, 231:289–337, 1933.

M. Nielsen and I. Chuang. *Quantum Computation and Quantum Information*. Cambridge University Press, 2000.

A. Peres. *Quantum Theory: Concepts and Methods*. Kluwer Academic Press, 2002.

I. Pitowsky. *Quantum Probability—Quantum Logic*. Springer, 1989.

B. Piwowarski and M. Lalmas. A quantum-based model for interactive information retrieval. In *Advances in Information Retrieval Theory*, volume 5766 of *Lecture Notes in Computer Science*, pages 224–231. Springer, 2009.

B. Piwowarski, I. Frommholz, M. Lalmas, and C. J. van Rijsbergen. What can quantum theory bring to information retrieval. In *Proceedings of CIKM*, pages 59–68. ACM, 2010.

J. Polkinghorne. *Quantum Theory: A Very Short Introduction*. Oxford University Press, 2002.

J. Ponte and W. Croft. A language modeling approach to information retrieval. In *Proceedings of SIGIR*, pages 275–281. Melbourne, Australia, 1998.

J. Renn. *Auf den Schultern von Riesen und Zwergen*. Wiley-VCH Verlag, 2006.

E. Rieffel and W. Polak. *Quantum Computing: A Gentle Introduction*. The MIT Press, 1st edition, 2011.

S. Robertson. The probability ranking principle in information retrieval. *Journal of Documentation*, 33(4):294–304, 1977.

S. Robertson and K. Sparck Jones. Relevance weighting of search terms. *Journal of the American Society for Information Science*, 27:129–146, 1976.

S. Robertson and S. Walker. Some simple effective approximations to the 2-Poisson model for probabilistic weighted retrieval. In *Proceedings of SIGIR*, pages 232–241, Dublin, Ireland, 1994.

S. Robertson and H. Zaragoza. The probabilistic relevance framework: BM25 and beyond. *Foundations and Trends in Information Retrieval*, 3(4):333–389, 2009.

J. J. Rocchio. Relevance feedback in information retrieval. In G. Salton, editor, *The SMART Retrieval System: Experiments in Automatic Document Processing*, chapter 14, pages 313–323. Prentice-Hall, 1971.

G. Salton. Associative document retrieval techniques using bibliographic information. *Journal of the ACM*, 10:440–457, 1963.

G. Salton. *Automatic Information Organization and Retrieval*. Mc Graw Hill, 1968.

G. Salton. *The SMART Retrieval System. Experiments in Automatic Document Processing*. Prentice-Hall, 1971.

G. Salton. Mathematics and information retrieval. *Journal of Documentation*, 35(1):1–29, 1979.

G. Salton. *Automatic Text Processing*. Addison-Wesley, 1989.

G. Salton and M. McGill. *Introduction to Modern Information Retrieval*. McGraw-Hill, 1983.

G. Salton, A. Wong, and C. Yang. A vector space model for automatic indexing. *Communications of the ACM*, 18(11):613–620, 1975.

K. Sparck Jones. *Automatic Keyword Classification*. Butterworths, 1971.

K. Sparck Jones and C. J. van Rijsbergen. Information retrieval test collections. *Journal of Documentation*, 32(1):59–75, 1976.

K. Sparck Jones and P. Willett. *Readings in Information Retrieval*. Morgan Kaufmann, 1997.

C. J. van Rijsbergen. *Information Retrieval*. Butterworths, second edition, 1979.

C. J. van Rijsbergen. A non-classical logic for Information Retrieval. *The Computer Journal*, 29 (6):481–485, 1986.

C. J. van Rijsbergen. *The Geometry of Information Retrieval*. Cambridge University Press, 2004.

V. N. Vapnik. *Statistical Learning Theory*. Wiley, 1998.

V. N. Vapnik. *The Nature of Statistical Learning Theory*. Springer, 1999.

J. von Neumann. *Mathematical Foundations of Quantum Mechanics*. Princeton University Press, 1955.

D. Widdows. Orthogonal negation in vector spaces for modelling word-meanings and document retrieval. In *Proceedings of the Annual Meeting on Association for Computational Linguistics*, pages 136–143, 2003.

D. Widdows. *Geometry and Meaning*. CSLI Publications, 2004.

D. Widdows and T. Cohen. The semantic vectors package: New algorithms and public tools for distributional semantics. In *IEEE International Conference on Semantic Computing (ICSC)*, pages 9–15, 2010.

D. Widdows and S. Peters. Word vectors and quantum logic: Experiments with negation and disjunction. In R. Oehrle and J. Rogers, editors, *Proceedings of Mathematics of Language*, volume 8, pages 141–154, 2003.

S. Wong and V. Raghavan. Vector space model of information retrieval—a reevaluation. In *Proceedings of SIGIR*, pages 167–185, Cambridge, UK, 1984.

W. K. Wootters. Statistical distance and Hilbert space. *Phys. Rev. D*, 23(2):357–362, 1981.

N. S. Yanofsky and M. A. Mannucci. *Quantum Computing for Computer Scientists*. Cambridge University press, 2008.

A. Zeilinger. *Dance of the Photons: From Einstein to Quantum Teleportation*. Farrar, Straus and Giroux, 2010.

C. Zhai. *Statistical Language Models for Information Retrieval: A Critical Review*. Foundations and Trends in Information Retrieval. Now Publishers Inc., 2008.

C. Zhai and J. Lafferty. A study of smoothing methods for language models applied to ad-hoc information retrieval. In *Proceedings of SIGIR*, pages 334–342, New Orleans, LA, USA, 2001.

G. Zuccon and L. Azzopardi. Using the quantum probability ranking principle to rank interdependent documents. In *Proceedings of ECIR*, Lecture Notes in Computer Science, pages 357–369. Springer, 2010.

G. Zuccon, L. Azzopardi, and C. J. van Rijsbergen. The quantum probability ranking principle for information retrieval. In *Proceedings of ICTIR*, pages 232–240, Toulouse, France, 2009. Springer.

Index

© Springer-Verlag Berlin Heidelberg 2015
M. Melucci, *Introduction to Information Retrieval and Quantum Mechanics*,
The Information Retrieval Series 35, DOI 10.1007/978-3-662-48313-8

Printed in the United States
By Bookmasters